Essential Strategies for Social Anxiety

Essential Strategies for Social Anxiety

Practical Techniques to Face Your Fears, Overcome Self-Doubt, and Thrive

Alison McKleroy, MA, LMFT

ROCKRIDGE PRESS

For general information on our other products and services or to obtain technical support, please contact our Customer Care Department within the U.S. at (866) 744-2665, or outside the U.S. at (510) 253-0500.

Rockridge Press publishes its books in a variety of electronic and print formats. Some content that appears in print may not be available in electronic books, and vice versa.

Interior and Cover Designer: Suzanne M. LaGasa
Editor: Lori Tenny
Production Editor: Ashley Polikoff

All images used under license © StatementGoods/Creative Market. Author photo courtesy of © Elisa Cicinelli Photography.

ISBN: Print 978-1-64611-930-1 | eBook 978-1-64611-931-8
R0

For my husband, Neil,

and for my parents.

Contents

Introduction

Hello and welcome! If you've come to this book because you want to permanently alter your experience of social anxiety and feel confident and free to be at ease in the world, then you've come to the right place.

Some people equate social anxiety with being shy or introverted, but there is more to the story. When you're socially anxious, it can feel like you're in the spotlight, with everyone judging you. It's hard to be yourself when you're feeling self-conscious, out of place, and afraid of being embarrassed. Friends and family may criticize you for avoiding social events, but it's natural to want to steer clear of the anxiety that can flood and overwhelm you.

I want to first assure you that you are not broken, and this book isn't about fixing something that's wrong with you. Like most people with social anxiety, you are likely just stuck and need effective tools to free and empower you so you can live your life with confidence and ease.

Learning new ways of being in the world is important work. Social anxiety doesn't just keep us from going to parties. When we let our fears and anxieties limit our range of participation in life, we diminish our joy and vitality. Our opportunities to experience connection and be seen and valued for who we are become restricted by self-doubt. It's painful to feel like we're on the outside looking in, longing for a sense of belonging. It's natural to feel lonely or alone.

No matter how long you've been restrained by anxiety, you can replace old patterns and learn how to bring ease and connection to your life in any moment, instead of waiting or hoping for it to happen to you someday. As a psychotherapist, trainer, and coach, I've helped over a thousand people learn and practice life-changing tools to transform their lives. I'm astonished by the results people achieve when they're empowered to face their fears and take action in the direction of what lights them up.

In fact, working with people who are struggling with social anxiety has been one of the most exhilarating parts of my professional career. The thrill and excitement that happens when people break through their fears and delight in their newfound freedom is contagious. It is the joy of my work. And I want you to experience the exhilarating transformation that I've been privileged to witness so many times.

Just like you can't learn to swim by reading a book, learning to overcome social anxiety will take more than gaining new insights. As you read these chapters, the real work begins when you complete the exercises and apply the techniques to your life. You don't have to wait until you have everything figured out to take action right away and start seeing results.

We have deep within ourselves—at every moment—the power and courage to change the quality of our lives. In fact, by choosing to pick up this book you've connected to it already. In each moment that you choose to be courageous over comfortable, to connect instead of retreat, you will experience an immediate difference in the confidence, freedom, and joy with which you live your life. I'm grateful and honored to guide you on this journey. Let's begin!

Author's note: The real-life examples cited in this book represent composites of actual clients. The names and identities have been changed to protect client confidentiality.

CHAPTER ONE

Social Anxiety and You

How important is social connection? Neuroscientist Matthew D. Lieberman, PhD, author of *Social: Why Our Brains Are Wired to Connect*, argues that our need to connect with others is as essential as our need for food and shelter. We all are wired to connect and belong, and our sense of community is one of the greatest predictors of our happiness.

If our well-being is so inextricably linked with our social connectedness, it's understandable that missing out on opportunities to experience connection feels so painful. When discomfort and anxiety prompt us to avoid social situations, the experience of being close to the world around us is obscured. If social fears are getting in the way of enjoying your life, you're not alone, and it is possible to have a breakthrough. You may doubt that you have an innate capacity to be open and connected. But you do!

In this chapter, you will start on the path to transforming social anxiety by looking at a range of stress-inducing social situations to see where you might be holding yourself back. You'll examine the socially anxious brain and how it impacts the body. You'll take a glimpse inside the anxious mind, home to the inner critic and perfectionist who love to point out faults and assume the worst about you. You'll get an overview of therapy and medication options and set some goals.

Social Anxiety Defined

My client Dave is yearning for a romantic partner. Yet when he meets someone he likes, Dave feels self-conscious and embarrassed, certain he's making a fool of himself. His heart beats intolerably fast, and he's sure everyone notices the trembling and blushing that broadcast his discomfort, despite his best efforts to hide it. The idea of meeting a woman at a party seems neither pleasant nor realistic, since he's been avoiding social gatherings for years. Dave is convinced he'll always appear awkward or out of place in any social setting, won't know what to say, or will come off as boring. He longs to share his life with someone, but he feels stuck and defeated.

It's perfectly normal for anyone to feel a little anxious in some social situations, like butterflies in your stomach before a date or your heart racing before a job interview. But for people like Dave, their anxiety related to social exchange is more than shyness, introversion, or nerves. Their interactions with others can cause such intense fear or stress that it impairs their social and professional functioning. Often their self-consciousness in social situations is so persistent, excessive, and uncontrollable that it greatly interferes with their daily life. Concerns about being judged, scrutinized, or rejected by other people can devastate their confidence and make everyday activities seem impossible.

Social anxiety exists on a continuum of severity, from shyness to feeling uncomfortable attending social engagements to living in fear of everyday conversations. Wherever you may have landed on the continuum, this book is for you.

Symptoms

If you're living with social anxiety, you'll recognize the symptoms of feeling self-conscious, embarrassed, or nervous in front of others. You might feel like a spotlight is on you, or that the people around you are evaluating every move you make. You expect the worst to happen in any social exchange, which leaves you feeling vulnerable and on edge. Afterward, you may beat yourself up over interactions that you believe could have gone better.

Some people with social anxiety experience some kind of disquiet in most, if not all, social situations. For others, the anxiety comes up in specific circumstances, like meeting new people, public speaking, or performing in front of an audience. You might have very specific most-feared scenarios, like speaking up in class or during a work meeting, making a phone call, or socializing at parties. A lot of people might find some of these situations challenging, but if you experience social anxiety, they can be so nerve-wracking that you'll go to great lengths to avoid them, like calling in sick or following a sudden urge to bolt from the room. You might anticipate such an event with great dread for days or weeks beforehand.

Social anxiety produces physical symptoms as well, and as we noted in Dave's story, they can be challenging to manage. You might blush, you might feel nauseous, your stomach might clench, your heart may beat fast, or your body might tremble. When you're talking to others, your voice might stammer or your mind goes blank. Some people feel dizzy, sweat, or have trouble breathing. The experience can feel intolerable, since along with the discomfort of the physical symptoms, you're also worried about how it must come across to everyone.

People with social anxiety typically understand that their anxiety is irrational, but knowing this doesn't do much to shift their persistent, negative thoughts. Your social anxiety can set off an unrelenting concern that other people think you're annoying, stupid, or look down on you. You might believe you're uninteresting, have nothing interesting to add to conversations, or that nobody wants to be around you. It can be exhausting to live your life with an unremitting fear that in any kind of social interaction your imperfections will become obvious to the people around you.

What Social Anxiety Is Not

If you have social anxiety, you might blame yourself for it and attribute your fears to a flaw in your character. I can assure you this is not the case. While the exact cause of social anxiety is unknown, experts believe both biological and environmental factors play a role. What's not clear is how much influence is connected to genetics and how much to learned behavior. Like most mental health issues, there are risk factors that contribute, including stressful life events like bullying or a family history of social anxiety. I often remind my clients that I don't know exactly how my car works, but that doesn't have to stop me from driving to where I want to go.

You might also be wondering if social anxiety has anything to do with general anxiety. Although the mental health profession does put social anxiety and general anxiety in the same diagnostic category, the two are distinctly different. Both are characterized by persistent and excessive worry related to a perceived threat. But generalized anxiety shows up as chronic anxiety or worry, while social anxiety involves excessive self-consciousness in social situations. In other words, social situations cause anxiousness even if other circumstances don't.

Is social anxiety just shyness? That question is one of the reasons that many people who have social anxiety wait a long time to seek help. They assume they're just shy and don't recognize the magnitude of the impact their condition has on their day-to-day lives. But while shyness and social anxiety share characteristics, what distinguishes social anxiety is the intensity of the fear it produces and the lengths someone will go to in order to avoid the possibility of being judged or rejected. For example, if you're shy, you may be uncomfortable going to a friend's birthday party but push yourself to attend and end up having a pretty good time. If you're struggling with social anxiety, however, you probably feel nervous just reading the invitation and will very likely turn it down.

Introversion is also a personality trait that's distinct from social anxiety. Introversion is a preference for low-level stimulation—a quiet evening at home instead of a night out at a club. An introvert invited to a friend's birthday might go for a little while and leave when they've had enough. Or they might choose not to go—not to avoid awkwardness and discomfort but because they're not in the mood for that level of stimulation or it might drain their energy.

If you have social anxiety, it's not unusual to feel hopeless and discouraged because you've been dealing with the problem for a long time without improvement. If you've been struggling to change things, don't worry. There are many effective tools you can learn to use and feel better right away, and with practice, you can permanently alter your social anxiety and connect with ease. We'll explore these proven strategies throughout the following chapters.

Anxiety-Inducing Social Situations

Most people who have social anxiety find their symptoms triggered by a variety of social situations. How many of these scenarios make you feel nervous or stressed?

Public performances, like playing a musical instrument in a concert, playing sports in front of spectators, or having to speak or give a presentation.

Workplace communication, such as talking to an authority figure, sharing your opinion in a meeting, or having to make phone calls in front of colleagues. All the steps required to get a job can trigger anxiety, from asking for a reference letter to the interview itself.

Classroom participation, like answering questions, presenting a report, or even entering the room when everyone is already seated.

Socializing, everything from making small talk, to introducing yourself to a group, to being in a room full of strangers. Going to a party, even with friends, can be nerve-wracking. Apprehension around dating includes fear of approaching someone you like and insecurity about keeping the conversation going on a date. Eating in front of others can be a trigger, making it hard to attend a dinner party or eat in a restaurant.

Sometimes just the anticipation of a trigger situation can be stressful, like receiving an invitation to your friend's wedding or being asked to a happy hour by your colleagues. When you can identify and understand the situations that trigger your anxiety, it's easier to learn ways to handle them. Consider the circumstances in which you're uncomfortable, and adapt the strategies in this book to create a breakthrough in those areas.

How Does Social Anxiety Affect You?

It can be hard to assess how big of an impact social anxiety is having on your life, especially if you've been struggling with it for some time. Broadly speaking, social anxiety tends to affect us in these ways:

Day-to-day life is difficult. When you're burdened by social anxiety, even small, mundane interactions at work, school, and elsewhere can cause fear and self-consciousness. It's hard to be around others when you're worried about embarrassing easily. You might blush or sweat, or your mind goes blank when forced to speak with someone you don't know. Simple tasks like getting a haircut or returning clothes that don't fit seem unachievable and are avoided as long as possible.

You miss out on opportunities. At work, you might never share your opinions for fear that colleagues will judge you as boring or obtuse, so you don't participate in discussions. You might be overdue for a raise, but the thought of talking with your boss is overwhelming and not worth the stress. If you're a student, lack of class participation might lower your grade, or you might drop a class if you're required to work in small groups. You might avoid restaurants and dinner parties, skip social gatherings, and not ask someone out, leaving you feeling isolated and lonely.

Your joy and connection are hijacked. It's painful to spend your days enduring such intense discomfort. So, of course, you stay away from situations that trigger your social anxiety. Why wouldn't you? Skipping them gives you relief, and your anxiety is reduced. Maybe you call in sick to work or change jobs. You may develop a habit of turning down invitations or cancelling plans with friends. You might seriously consider not showing up to your brother's wedding or your mother's retirement party. The problem is that while avoiding your triggers brings short-term relief, in the long term it keeps anxiety alive. It can be lonely missing out on opportunities to connect with others. How has avoidance impacted your relationships and your sense of connection? Are you wanting to feel close to friends, have fun with colleagues, or grow your social circle?

The strategies in the following chapters will equip you to handle social situations instead of avoiding them and empower you to create connection and experience the joy of belonging.

SOCIAL ANXIETY SELF-ASSESSMENT

As we've discussed, many social situations can trigger anxiety. This assessment tool will help you determine where social anxiety is showing up in your life. (Note: Designate a notebook or journal to use for the exercises throughout this book. It's a good idea to have all your work together so you can track your progress, return to previous exercises, and review your responses.)

Read each scenario and rate the level of anxiety you feel in each situation: None, Mild, Moderate, or High.

- Talking to someone you don't know.
- Asking someone out or going on a date.
- Talking to a friend over lunch.
- Giving a presentation in school or a meeting at work.
- Asking a salesperson for help while shopping.
- Performing in front of an audience.
- Public speaking.
- Making small talk when there's a lull in conversation.
- Having friends sing "Happy Birthday" to you.
- Eating in public.
- Participating in a small group.
- Talking to an authority figure.
- Mingling at a social gathering.
- Sharing your thoughts in a meeting.

Are there others? Add as many to the list as you can think of.

It can be helpful to have a focus on the areas where you'd most like to see a change. I invite you to look at the situations that you designated with moderate or high anxiety levels and choose up to three that you'd like to work on. As you move through the exercises and strategies in this book, think about how you can apply them to these situations in particular. We'll be referring back to this exercise at the end of this chapter.

The Brain-Body Connection

Anxiety may not be pleasant, but it's a natural, and even purposeful, emotion for human beings. In fact, having some anxiety keeps you safe: A sense of fear and apprehension can make you alert and ready to deal with any dangerous threats coming your way. It also motivates you to take action to solve whatever problem is causing the anxiety. Anxiety is part of our body's survival mechanism, allowing us to react quickly to dangerous situations.

To understand social anxiety, let's take a closer look at the body's stress reaction, also known as the fight-or-flight response. When a threat is detected, an almond-shaped structure in your brain called the *amygdala* is activated and sends a message that triggers the release of stress hormones. These chemicals flood your system and prepare your body for action. Your senses and perception are heightened, your heart beats faster, and your reflexes are quicker. Your body gets an energy boost, fueling your muscles so you can better fight or run away.

You can see how that stress reaction is a survival mechanism, allowing your body to turn its full attention to handling a dangerous situation. When the threat passes, the stress response turns off so your body can go back to normal. The fight-or-flight reaction is replaced by what's called the rest-and-digest state. This produces a calm and relaxed feeling, slows things down, and undoes the work of those stress hormones.

Our body's stress response is supposed to engage when needed, and then disengage after. Unfortunately, today's complicated world sometimes keeps us alert and on edge all of the time, day after day. It's as if the fight-or-flight switch is stuck in the "on" position. Our threat-detection system, great at scanning for a tiger that might be hiding in the jungle, instead becomes triggered by our anxious and negative thoughts. Research suggests that the brains of people with social anxiety might be more prone to interpret social interactions as real, physical threats. This activates the stress response system and triggers a cascade of physical reactions that aren't very helpful when you're trying to invite someone on a date or play piano on stage at a concert.

Physical Manifestations of Social Anxiety

When you have social anxiety, the world can seem like a hazardous place, filled with judgmental people who can attack at any moment with their scrutiny. Instead of scanning for a tiger, your instincts are surveying the environment for social situations that might involve negative judgment. If you consider that your brain perceives people with negative opinions as if they were dangerous animals, it's easy to understand why social anxiety produces the symptoms it does. As your body prepares to fight or run from a threat, your breathing becomes rapid and shallow, bringing in more oxygen. Your heart rate speeds up to circulate blood to your muscles, which can make your chest feel tight. Some blood vessels widen to handle the increased blood flow, causing your face to flush. You sweat to control the heat generated by your overactive circulatory system.

Meanwhile, processes in your body that aren't directly related to coping with danger are blocked. This includes digestion, which is why you might experience appetite loss, nausea, diarrhea, or an upset stomach. And you know what else your body considers "nonessential" in an emergency? Complex thinking! The part of your brain involved in planning, concentration, memory, and abstract thinking shuts off. This is why you might experience your mind going blank, or find yourself stumbling over your words and struggling to remember what you were going to say.

Over time, social anxiety isn't just bad for our social lives, it can take a toll on our physical health. Our bodies have an amazing capacity to deal with stress in the short term. But chronic or excessive stress, worry, or anxiety can wreak havoc on our health and well-being. Staying in that fight-or-flight mode for long periods overexposes us to stress hormones that can trigger a host of health problems, from irritable bowel system (IBS) to heart disease.

But there's good news: There are many tools that can manage and shut down the stress response, activating our restorative, rest-and-digest system in its place. Practiced regularly, relaxation and mindfulness techniques, for example, can counteract the effects of stress both in the short term and the long term. We'll explore these techniques later in this book. We'll also discuss how you can untwist your anxious thinking, preventing it from triggering your body's stress response in the first place.

The Anxious Mind

From an evolutionary perspective, we can see why it's not a great idea for human beings to stay relaxed when faced with a threat. Scanning for danger and reacting to threats is crucial to self-preservation. When you struggle with social anxiety, though, it's no longer a crisis situation setting off the alarm bells. Rather, it's your anxious thoughts about being scrutinized by other people that trigger the body's stress response. Let's look at some common ways that those anxious thoughts express themselves.

The Inner Critic

Most people with social anxiety worry that they will do or say the wrong thing, and that they'll embarrass themselves while interacting with others. You may be concerned that some of your deficiencies will be exposed in a social interaction, and you'll be rejected for them. It can feel like there's a harsh inner critic inside your head, a voice that's continually highlighting all of your flaws. Your inner critic might shout, "You're not interesting enough, and you're putting them to sleep! You'll look awkward, and everyone will notice you blushing. You won't know what to say and you'll make a fool of yourself!"

Sound familiar? Those are just some of the many critical thoughts you might have when your social anxiety is active. We all have an inner critic that likes to point out our faults, tries to read other people's minds, and tells us when we're not living up to our expectations. To some degree, it's helpful to evaluate our mistakes. But when the inner critic becomes loud and contentious, we can feel pretty bad about ourselves. Over time, we come to take our inner critic's running commentary as fact and get stuck with persistent self-doubt.

Becoming aware of this inner critic is the first step to loosening its vice-like grip. Once we're able to recognize the negative thoughts we're telling ourselves (and stop believing them to be true), it's easier to unhook from the limited stories we've created. As we work together to free you from social anxiety, we'll explore ways you can quiet your inner critic and develop a more realistic and compassionate evaluation of yourself.

The Perfectionism Pit

Often our inner critic—and all the faults it points out—is a reflection of the high standards we have for ourselves, and how well we believe a social situation "ought to" go. Our inner critic can demand perfection and accept nothing less. With social anxiety, a big fear is someone finding out that you have defects and thinking less of you because of it. That being the case, it makes sense to try to be perfect. Only by ensuring that none of your imperfections are revealed can you avoid negative judgment and shame.

Some examples of perfectionist thinking include, "I can't make a mistake in this meeting. Everyone will think I'm stupid and think less of me." Or "I should never appear nervous or weak." Sometimes this thinking can create a habit of overpreparing, like spending an hour rewriting a brief email to your boss to ensure that it's flawless.

If you believe your interactions with others should be impeccable and you can't make a mistake, that kind of pressure will naturally leave you feeling nervous every time you have to interact with someone. While it can be helpful and motivating to have high standards, there is a distinctive difference between the pursuit of excellence and needing to be flawless to avoid failure and shame.

And there are other costs to perfectionism. Striving to be perfect in social situations is an exhausting gambit. When you focus exclusively on obscuring your shortcomings, you miss out on possibilities for authentic connection. For example, if you spend a meal with a colleague monitoring what you share (or don't share) to minimize the pain of judgment, you pass up an opportunity to be seen and accepted for who you are.

In her book *The Gifts of Imperfection*, Brené Brown, PhD, research professor, international speaker, and author, describes perfectionism as "a twenty-ton shield that we lug around thinking it will protect us, when in fact it's the thing that's really preventing us from flight."

All humans make mistakes. It's impossible to learn anything without making some errors along the way. Mistakes can be incredibly helpful. Consider them as precious life lessons that teach us how we want to do things differently next time. Dropping your protective shield takes courage but can offer you the possibility of being free, connected, and seen for your true self.

The Avoidance Trap

Avoidance and escape are natural reactions when we're dealing with social anxiety. We're hardwired to avoid anything threatening. Who wants to lean into discomfort and awkwardness when there's a way to get around it?

But avoidance is not the solution for social anxiety that it may seem to be. Let's consider the story of John, one of my clients, who became skilled at avoiding anxiety-provoking social situations. Whenever he was invited to a party, he declined with a great excuse. When he did go out to dinner with friends, John always found a way to duck out early, sometimes before the main course arrived. As soon as he said "no" to an invitation or left the restaurant, his anxiety went away . . . for a while.

Avoidance is a form of negative reinforcement—it rewards you for not participating. John's avoidance of social situations was rewarding in the short term, because he escaped the discomfort of speaking with people he didn't know and the potential pain of being rejected. His strategy worked successfully almost every time, so he kept it up.

The problem with avoidance is that it simply doesn't work in the long run. When we avoid socializing, we never develop the skills to do anything but keep avoiding it.

Over time, John became more fearful of social gatherings. He felt lonelier and more isolated, and his anxiety became worse. He came to me for help because despite his short-term gratification, John got tired of being stuck in a pattern of choosing comfort over courage. He felt that he was not living the life he imagined for himself. John wanted to be accepted and liked, and he longed for more fun in his life.

Playing It Safe

Another unhelpful way that people cope with social anxiety is engaging in *safety behaviors*. These are behaviors used to try to reduce or mask being embarrassed or self-conscious. They are an attempt to control your situation and manage your fears.

Whenever my client Jin ate dinner with friends, she usually covered her face with a napkin in case she spilled food or looked weird while she ate. She talked softly and rehearsed what she was going to say before the meal. When she wasn't eating, Jin would keep her hands in her pockets so no one could see if they were shaking when she was nervous. Throughout dinner, she would avoid eye contact so she wouldn't be noticed or feel overwhelmed. She tended to ask the person sitting next to her endless questions to keep the focus off herself. At the end of the meal, you would undoubtedly find Jin in the kitchen doing all of the dishes to avoid any further interactions.

Safety behaviors can be go-to moves that you rely on in a specific situation, like wearing headphones on public transportation or not looking up while giving a presentation. They can include scrolling through your phone, avoiding eye contact during conversations, showing up exactly on time for meetings to avoid small talk beforehand, or arriving early to avoid entering a room where others are already seated. Safety behaviors can also be a spur-of-the-moment idea to cope with discomfort, like pretending to fall asleep on an airplane.

The problem with safety behaviors is that, like avoidance, they momentarily get rid of your anxiety, but it returns the next time you're in the same situation. Both measures not only don't reduce your anxiety in the long run, they actually *maintain* your anxiety. They prevent you from giving yourself a chance to develop the skills to manage the situation. And when you avoid or play it safe, you never get to experience firsthand that you can actually handle it.

Another drawback of safety behaviors is that they can produce the *opposite* outcome from the one you wanted. When Jin keeps her napkin over her face while she eats, she actually appears more awkward. Or when she asks a barrage of questions to avoid having attention on herself, the other person can become annoyed or think Jin is being strangely interrogative.

Furthermore, some safety behaviors can be harmful to your health. Using alcohol or drugs as a way to cope, for example, can lead to more severe problems, like substance abuse or dependence.

Flipping the Script

The key to breaking through your social anxiety is to challenge your negative thoughts and face your anxiety-provoking situations without the prop of avoidance or safety behaviors. It takes vulnerability and a willingness to be present and allow feelings of discomfort to pass through you, but the rewards are immeasurable. When you become disentangled from the trap of avoidance and safety behaviors, you are free.

I've seen these breakthroughs happen again and again. It took work for John to stop believing his negative thoughts. Getting to the point where he could regularly attend social gatherings without cancelling or playing it safe was a gradual process. John's results didn't happen overnight, but he now deeply enjoys parties and is dating a woman he loves.

Jin started her process by having dinner with one friend, not hiding behind a napkin or avoiding eye contact. Over time, she worked her way up to hosting a dinner party for six people for her birthday. And when she spilled a crumb of cake on her shirt, she knew what positive thoughts to tell herself, and got to experience firsthand that no one cared or even noticed! She now feels completely at ease during meals shared with other people.

Standing Up to Social Anxiety

A glance at this book's table of contents will tell you that there are many proven solutions that can help you overcome social anxiety. In the following chapters we'll take a deep dive into each of the following strategies in turn, discussing along the way numerous techniques and real-world scenarios, so you can powerfully and confidently transform the anxiety-inducing social situations in your life. Feeling confident and at ease is within your reach. Here's a preview of the different approaches that we'll be exploring.

Cognitive Behavioral Therapy

Cognitive behavioral therapy (CBT) is a hands-on, practical approach to problem-solving and is based on the idea that the way we think determines how we feel and what actions we take. For example, if you believe that everyone is judging you, you'll most likely feel anxious and self-conscious, and try to get out of social situations in which you might be judged.

In chapter 2, you'll learn that CBT offers strategies to recognize and challenge detrimental patterns of thinking, known as *thinking errors*, such as polarized thinking, fortune-telling, or overgeneralizing. The goal is to help you replace negative thought patterns that are causing you distress with more accurate, positive thoughts that improve mood and behavior as well as your outlook on life.

Exposure Therapy

Exposure therapy is a treatment to help people safely confront their fears and break the pattern of avoidance. Chapter 3 discusses systematic techniques with which you can gradually expose yourself to various aspects of your fear, ultimately reaching the point where you feel confident you can handle it. The most common method is *graded exposure*, in which you rank feared situations in order of severity, and gradually expose yourself to each step, beginning with the mildest. Other options include *flooding*, which is exposure to the most-feared situation, and *systematic desensitization*, which uses relaxation techniques to reduce your physiological fear response.

Acceptance and Commitment Therapy

Chapter 4 explores a different framework for handling social anxiety: acceptance and commitment therapy (ACT), an action-oriented therapeutic approach. ACT is based on the belief that human suffering is natural and inevitable, therefore trying to suppress or avoid negative feelings is not only ineffective but actually ushers in more distress. When using ACT techniques, you commit to embracing your negative thoughts, feelings, and body sensations rather than trying to suppress what causes you discomfort. Self-acceptance and mindfulness skills help you to deal with social challenges that come your way.

Mindfulness and Meditation

I've mentioned mindfulness a few times in this chapter, and in chapter 5 we'll take a closer look. Mindfulness is the ability to bring full awareness to where you are and what you're experiencing in the here and now, without judgment. This helps with anxiety because we can't be caught up in our negative stories about ourselves and be in the present moment at the same time. The concept originated in Buddhism and was made more accessible in the Unites States in a program called mindfulness-based stress reduction (MBSR). Learning to meditate is one practice of mindfulness that can grow your capacity for being calm and present, the antidote to being anxious and fearful.

Communication Skills

In chapter 6, you'll supplement the strategies you've learned for managing social anxiety with an assortment of new, practical skills that you can employ in various social situations. Working on these skills is crucial to being able to effectively handle social interactions with ease and confidence. We'll cover assertiveness, active listening, nonverbal communication, and a range of other communication techniques.

Medications and Individual Therapy

Sometimes a combination of working with a therapist and taking medication can be useful for treating social anxiety. Several different types of medications are prescribed to treat social anxiety. Each medication has its advantages and disadvantages. Here are three categories of medication that a medical professional might discuss with you. (Note that only certain health care providers, like your doctor or psychiatrist, can prescribe medications.)

SSRIs

Selective serotonin reuptake inhibitors (SSRIs) are considered the first choice for general symptoms of social anxiety disorder. Common SSRIs include Paxil (Paroxetine), Zoloft (Sertraline), Prozac (Fluoxetine), and Luvox (Fluvoxamine). These medications cause a brain chemical called serotonin to last longer. Serotonin assists in mood regulation; when serotonin levels are stabilized, feelings of anxiety are decreased. Possible side effects of these medications are fatigue, insomnia, and drowsiness. They can also cause irritability, weight gain or loss, and decreased sex drive.

SNRIs

Serotonin and norepinephrine reuptake inhibitors (SNRIs) target not only serotonin but also norepinephrine (adrenaline), a chemical that regulates our fear and stress response. Common medications include Effexor (venlafaxine), Cymbalta (duloxetine), and Pristiq (desvenlafaxine). SNRIs increase your levels of both serotonin and norepinephrine, which improves mood while reducing anxiety and panic attacks. SNRIs can have more side effects than SSRIs.

Possible side effects include nausea, dry mouth, fatigue or drowsiness, headaches, and even increased nervousness, especially at first. People also report sleep problems and sexual response difficulties. You should avoid drinking alcohol if you are taking SNRIs; alcohol can interact with SNRIs to cause increased intoxication and drowsiness.

Benzodiazepines

Benzodiazepines are sometimes used on an as-needed basis for people who experience panic attacks or have specific social phobias. These are muscle relaxants that act as a sedative; they work by slowing down the central nervous system, which can relieve anxiety and induce relaxation. They can be taken before a specific situation that triggers anxiety, such as before giving a performance or speech. Common medications include alprazolam (Xanax), clonazepam (Klonopin), diazepam (Valium), and lorazepam (Ativan). Drowsiness and dizziness are the most common side effects. Unfortunately, benzodiazepines can be habit-forming and lead to physical dependency, especially if taken for more than two weeks. For this reason, doctors use caution when prescribing benzodiazepines to people with a history of substance abuse.

Individual Therapy

Social anxiety can cause a lot of distress and disrupt your ability to participate fully in your life. While you will find dozens of actionable strategies you can use on your own in this book, I urge you to see a therapist trained in social anxiety as part of your game plan. From helping you untwist your negative thoughts, to creating an exposure hierarchy (we'll discuss this in chapter 3), to practicing social skills with you in a role-play scenario, personalized guidance from a professional can be an invaluable part of the process. It can be tailored to your specific, individual needs. Treatment with a therapist can help you to create a personal plan and stay accountable as you build the skills to address your fears and concerns. I encourage you to look for a therapist who is certified in CBT and/or ACT, trained in exposure treatment, and uses a mindfulness approach. Most therapists offer free phone consultations to make sure it's a good match before starting therapy together. You'll find some guidance for seeing a therapist in the Resources section at the end of this book.

GOAL SETTING

Now that you've had a chance to identify your social anxiety triggers and see the ways that social anxiety can interfere with your life, I invite you to develop some goals you would like to work toward that will guide and inspire your efforts. When you have a clear vision of how you'd like your life to look, it will be easier to stay focused and make the changes needed.

How to do it:

Review your self-assessment quiz (page 7), and note which situations cause you moderate to high levels of anxiety. From those situations, choose three for which you'd like to overcome your anxiety, and use this template to develop a goal related to each. Write down your answers in your notebook or journal, which you can refer to as you continue to work through this book.

1. Describe the situation that triggers your anxiety.

   Examples: *Attending social events, eating in front of others, talking to strangers, speaking in public, going on a date*

2. Describe the physical symptoms you typically experience.

   Examples: *Shortness of breath, heart racing, sweating*

3. Describe the negative thoughts you typically have.

   Examples: *I'll look awkward, I'm not interesting enough, I'll say something embarrassing*

4. What are your avoidance, escape, or safety behaviors?

   Examples: *I turn down invitations, I avoid eye contact, I leave early, I eat with a napkin over my mouth, I make phone calls from the bathroom*

5. Choose a goal by finishing the following statement.
   If I had a magic wand and my social anxiety suddenly disappeared, I would be different in this social situation in the following ways:

   Examples: *I would say yes to invitations, speak up in meetings or in class, stay at the party as late as I want to, ask someone out on a date, or go out to lunch with colleagues*

How to Use This Book

The strategies in this book are evidenced-based and proven to be effective through research. Proceed at your own pace, and I recommend working through the book from start to finish, as the information in the chapters build on each other. It's best to complete the strategies in each chapter before moving to the next. Don't forget to designate a notebook or journal to be used exclusively for the exercises throughout the book. And remember, this book is meant to guide you to take action. It takes more than knowledge and insight to shift social anxiety, so you'll need to practice the techniques to generate new results for your life.

The journey to overcoming social anxiety can be, at different times, scary, exciting, surprising, and wondrous. Some parts of the process may be harder or take a little longer than others. But you are already on the path and have been since the moment you picked up this book. By implementing the various strategies and techniques we'll be discussing, you'll experience how possible it is to grow your courage and confidence, and make positive, permanent changes in the quality of your life.

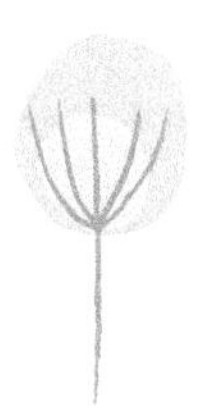

CHAPTER TWO

Think Different, Be Different

Imagine you're invited to a happy hour by a few coworkers you like. You want to go, but in the past, you haven't felt comfortable at these kinds of things. If you tell yourself, "I'll seem nervous and everyone will notice," you'll likely feel anxious and decline the invitation. On the other hand, if you tell yourself, "It's okay to feel nervous. It's part of being human. I can do hard things and get through this," you'll probably feel more confident and willing to go. The way you think influences both how you feel and what actions you take.

And that's the concept behind cognitive behavioral therapy (CBT), considered one of the most effective approaches for treating social anxiety. The power of CBT lies in its emphasis on how distressing thoughts cause our pain and discomfort. Our beliefs about ourselves, the world, and the future are what determine our mood and behavior, not our circumstances.

In this chapter, we'll explore what CBT is, how it was developed, and how to apply powerful strategies to free yourself from anxious thinking patterns triggered by social situations. When you're able to identify and challenge the problematic thinking that causes your anxiety and distress, you start to feel better and gain new freedom.

CBT: A Primer

"The mind that opens to a new idea never returns
to its original size." —Albert Einstein

Back in the 1960s, psychiatrist Aaron T. Beck, MD, and psychologist Albert Ellis, working independently, noticed that their patients had automatic (that is, instantaneous and habitual), negative thoughts and irrational beliefs about themselves, the world, and the future. When these negative interpretations went unchallenged, they impacted their patients' moods and behaviors without them realizing it. But when they were able to discriminate between their own thoughts and reality, they felt better and acted differently.

Dr. Beck developed this approach into a treatment method called cognitive behavior therapy (CBT). As the name suggests, its aim is to change both cognition (your patterns of thinking) and behavior (your actions). CBT is a goal-focused therapy, offering a practical, hands-on approach to problem-solving. This is achieved through strategies that you'll learn about in this chapter, including thought logs, behavioral experiments, and other methods designed to challenge your existing thought and behavior patterns.

Dr. Beck's theory of cognitive therapy was innovative in its emphasis on the role of thinking and its influence on our emotional distress. Central to CBT is the concept that it isn't our circumstances that cause us suffering but our *interpretation* of our circumstances or reality. In other words, what's happening in the world around us doesn't determine our emotional well-being. The stories we tell ourselves about what's happening do.

Here's how that plays out. Let's say you see some of your coworkers chatting in the break room at work, and you immediately think, "I'm dull and they don't want to be around me." This automatic negative thought causes you to feel a certain way in the moment (nervous, lonely, self-conscious) and guides what you do next (walk past the room, avoid eye contact). A group of coworkers talking and laughing is not what's causing you to suffer. What's causing your pain is the story you are making up about yourself—and believing.

And this is actually good news! There's not a lot we can do about how other people behave, and we don't have much control over the challenges life throws at us. But when we see that our suffering is caused by our own negative, distorted thoughts, which keep us stuck in a state of anxiety, depression, or other painful states, we're empowered to get unstuck.

That focus on our thoughts is what makes CBT so effective, especially for treating social anxiety. If you walk by your coworkers and can distinguish what's real (a group of people talking to each other) from the negative conclusions you are jumping to ("They don't want to be around me"), you create space for yourself to feel and act differently. In this example, you could remind yourself, "I don't have concrete evidence for what they think about me, and their approval doesn't determine my worth as a human being anyway." With that in mind, you might feel more confident and take a risk in joining their conversation. When you're able to uncover problematic thoughts in this way, challenge them, and replace them with ones that are positive and rational, your experience of yourself and the world drastically changes. And this can happen right there in the moment.

Because it is the most rigorously researched form of psychotherapy, CBT is often considered the gold standard in therapy for treating a range of issues, including anxiety, depression, phobias, insomnia, obsessive compulsive disorder (OCD), and more. CBT's methods are founded on scientific studies and evidence that support its claims. Having worked with hundreds of clients using CBT in my practice, I can confidently attest that it works! The huge transformation that occurs between someone's first session and their last is a remarkable testament to how effective the tools of this therapy can be. Many clients complete their therapy radiating so much confidence and courage that they're almost unrecognizable. You can experience the same changes in your life as you learn to understand, manage, and overcome your social anxiety.

An advantage of CBT is that it helps reduce symptoms right away. The moment you no longer believe a thought—"I looked like an idiot in the meeting"—and replace it with different, more credible ones—"I can't read people's minds . . . My mind went blank because I was nervous, which happens to everyone . . . Feeling like an idiot doesn't mean I *am* an idiot"—you'll immediately feel and act differently. And the strategies you learn to untwist your negative thinking will be available to you whenever you need them, for the rest of your life. It's like discovering balance when you learn to ride a bike. Once you've mastered that balance, you'll always be able to ride a bike, regardless of how long it's been sitting in your shed.

While CBT is similar to other therapies, its distinctive approach is guided by the following core principles.

Collaboration and Active Participation

CBT is a collaborative effort between therapist and client. Cognitive behavioral therapists don't tell their clients what to do. Instead, they work together collaboratively to solve problems. Clients take an active role, including planning and completing homework assignments between therapy sessions. A cognitive behavioral therapist helps you help yourself in the areas that matter most to you.

A Goal-Oriented Focus

Cognitive behavioral therapy is a problem-solving therapy focused on helping you achieve your specific goals. You begin treatment with the end in mind. You create objectives that are relevant to you, with input from your therapist. As you experienced at the end of chapter 1, creating goals is in itself a valuable process, because it requires you to take charge of your problems and their solutions.

Emphasis on the Here and Now

Focusing on dealing with your current struggles rather than the cause of your problems is an integral part of CBT. Understanding why you have social anxiety can be interesting, but knowing why won't solve the problem. It's changing thoughts that are negatively impacting you in the moment that empowers you to take new actions to produce different results.

Being Your Own Therapist

A key distinction between cognitive behavioral therapy and other treatments is that with CBT, you learn to treat yourself. You're trained to be your own therapist, so you can use tools to untwist your negative thinking on your own, outside of the therapy sessions. You learn techniques to work with your own symptoms. And you continue doing this after treatment is over, without further help from a therapist.

Relapse Prevention

Learning how to feel better—and keep it that way—is an important part of cognitive behavioral therapy. It's natural to fear losing the progress you made and to be concerned about slipping back into old ways of thinking and acting. Relapse prevention techniques help you keep on the lookout for triggers that could get you off track, and develop strategies to cope with them.

Time-Limited

CBT is significantly shorter in duration than traditional talk therapy, which tends to be more open-ended with no clear end date. With CBT, once you've met your goals, treatment usually ends. Many people can complete treatment for social anxiety in a few months to a half-year, though more severe cases may take longer to resolve. Some people resolve their social anxiety very quickly, depending on how willing and eager they are to face their fears. If you're willing to take big risks, you can even overcome your social anxiety in one moment of powerful self-discovery or a bold step in confronting your fears.

Structured

Each CBT session has a predetermined structure to ensure that therapy appointments don't become chat sessions. Home-practice assignments allow significant improvement to happen quickly.

Addresses Automatic Negative Thoughts

Negative automatic thoughts (NATs) are negative interpretations of ourselves and what's happening in the world around us. They're self-defeating and negatively impact our mood and confidence. We tend to automatically believe our negative thoughts and don't pause to question their validity. When you apply CBT techniques, you can recognize what your mind is automatically telling you, then test those thoughts for accuracy so you can develop more helpful ways of thinking.

A Variety of Techniques

Cognitive behavioral therapy includes a number of different evidence-based methods to change thoughts, feelings, and behaviors, and improve your mood and overall life satisfaction. One of the most commonly used methods is called *cognitive restructuring*, which helps you identify and shift thinking patterns responsible for your negative mood and ineffective behavior. We're going to take a deeper look at that technique right now, and in upcoming chapters I'll introduce other tried-and-true practices for overcoming social anxiety, including graded exposure, mindfulness, and skills training.

Your Thoughts, Reimagined

The way we think affects the way we feel and what we do. From this central premise of cognitive behavioral therapy, it makes sense that distorted interpretations about ourselves and the world—that is, inaccurate thinking—will cause us problems. The skills learned in CBT aim to reduce social anxiety and other negative feelings by testing unhelpful thoughts for their accuracy, and replacing inaccurate, negative thinking patterns with more positive and functional ways of thinking. That process is called cognitive restructuring. You might think of it as rewiring your own brain, creating new mental pathways to override unhealthy thought patterns. I often tell my clients that their anxious thinking patterns are like a well-trodden path or even a slide, while cognitive restructuring is like taking a machete and cutting down bushes to clear a new trail. It takes a lot of effort in the beginning, but eventually, you've created another route to take.

Cognitive restructuring equips you with many options for unraveling your negative thinking. Two strategies that you can learn right now are described by world-renowned psychiatrist and Stanford University Professor Emeritus David D. Burns, MD, in his book *When Panic Attacks: The New, Drug-Free Anxiety Therapy That Can Change Your Life*. He explains how a negative thought can be defeated through acceptance or self-defense—or by using both.

For example, suppose you tell yourself, "I'm a failure." One way to transform that negative, irrational thought is to accept some truth there may be to it with a sense of peace and acknowledge that truth with a more accurate statement: "I do sometimes fail and make mistakes." To take a more defensive stance against the thought, you could assert why it's incorrect: "Just because I make mistakes doesn't mean I'm a failure. I can't learn without making mistakes." And a combination of both techniques might sound like this: "While it's true I do sometimes make mistakes, it doesn't mean I'm a failure. It just means I'm human and I can learn from my mistakes."

A common misconception about cognitive restructuring is that it's just a form of positive thinking. But we're not talking about creating a positive mantra for you to repeat in the mirror and try your best to believe. With cognitive restructuring, the goal is not just to think more positively, it's also to think more realistically. In order for a new, positive thought to be effective, it has to be completely believable and, most importantly, it has to defeat the believability of the negative thought. Let's take a closer look at cognitive restructuring in action.

Maya finally gave a presentation at work that she'd been dreading for months. When she finished, her colleagues and manager praised her for her hard work and pointed out what they liked about it. But when her manager Hillary added a small criticism about the content of one of her slides, Maya's heart sank. She felt humiliated and defeated, and she could feel her cheeks turning red. Maya anxiously counted the minutes until the meeting was over.

Back in her office, Maya began engaging the CBT strategies she'd learned. First, she completed a thought log and recorded some of her negative feelings: unhappy, worried, embarrassed, and defeated. Next, she wrote down some of her negative thoughts: "I completely screwed up," "Hillary thinks less of me now," and "I'm going to get fired." Next, she identified some of the thinking errors in those thoughts: She was mind-reading by assuming what other people were thinking and ignoring the good by not acknowledging the praise she'd received. She was filtering by focusing only on Hillary's one piece of criticism and catastrophizing by predicting the worst possible outcome. (See the chart on page 28 for more about thinking errors.)

Maya reviewed what she wrote and began challenging the validity of these thoughts using a variety of CBT methods, including examining the evidence for and against her negative thoughts, and decatastrophizing. She was eventually able to reframe her thoughts with more accuracy. Her counter-statements aren't blindly positive, but they put the negatives in perspective. "One incorrect slide doesn't mean I screwed up the whole presentation and isn't a serious enough offense to get fired over... Hillary also praised my presentation and has given me two positive performance reviews in the last two years... Even if I were fired, I could look for another job." In the past, a situation like this would have sent Maya into a downward spiral of dejection and panic, but she's learned how to get herself unstuck from her negative thoughts and feel better right away.

Does it seem impossible to change the way you think in social situations? It's true that negative thinking, like the worry and rumination you might have about a social encounter, can be a strong habit. It can even seem normal when you've been doing it for so long. But think of other challenges you've had in your life: learning to drive a car or play a musical instrument. Cognitive restructuring is like any other skill that takes focused work and attention. With practice, you can learn to stop automatically trusting your negative thoughts as true and real. It's not only possible, it can even be exciting to free yourself from the burden of believing your negative self-talk.

The Mind Games We Play

Thinking errors, or cognitive distortions, are biased perspectives we have about ourselves, our situations, other people, and the future. They are negative, irrational thoughts and beliefs that we unconsciously reinforce.

Thinking errors were first described by Aaron T. Beck, the founder of CBT, in his book *Cognitive Therapy of Depression*, published in 1979. David D. Burns, a psychiatrist who also contributed to the birth and evolution of CBT, continued Beck's research and popularized 10 of the most common cognitive distortions in his best-selling book *Feeling Good: The New Mood Therapy*, first published in 1980.

Since Beck and Burns, new thinking errors continue to be identified by researchers. There are many ways our thinking can be biased, and Wikipedia even keeps an ongoing list of more than 100 cognitive biases.

	THINKING ERROR	DESCRIPTION	EXAMPLE
1	Polarized Thinking	Black-and-white thinking. You view things in extremes, like either/or, good/bad, right/wrong, or all/nothing. You fail to see the middle ground.	"Unless everyone in the meeting likes my presentation, it's a total failure."

	THINKING ERROR	DESCRIPTION	EXAMPLE
2	Personalizing	You tend to take things personally. You blame yourself for circumstances outside your control.	"My teacher's annoyed facial expression must be because of something I did."
3	"Shoulding"	You impose unrealistic expectations on yourself, others, and the world based on how you expect life should be. You set the bar too high.	"I shouldn't be so ill-prepared. I should have known exactly what to say when she asked me about my summer vacation plans."
4	Overgeneralizing	You make hasty generalizations from insufficient evidence. You draw broad conclusions from a single negative event and expect it to happen again and again. Overgeneralizing thoughts often include the words "always," "never," "every," or "all."	"Coffee dates never work out."
5	Filtering	You focus on the negative details of a situation and filter out any positive aspects.	"That person yawning in the audience means my performance was boring, no matter what my friends tell me."

	THINKING ERROR	DESCRIPTION	EXAMPLE
6	Ignoring the Good	You reject positive experiences or your positive qualities by insisting they don't qualify.	"It was a fluke she said 'yes' when I asked her out. She's probably just desperate for company."
7	Mind-Reading	You believe you know what other people are thinking about you without sufficient evidence.	"Janet's friend thinks I'm weird and awkward."
8	Fortune-Telling	You make negative predictions about the future based on little to no evidence.	"My presentation is going to go horribly this week."
9	Spotlight Effect	You believe you are being noticed more than you really are. You overestimate your effect on others.	"Everyone at dinner noticed me spill a crumb on my shirt."
10	Negative Labeling	You attribute your actions to a flaw in your character. Instead of assuming you made a mistake, you assign a negative label to yourself.	"I'm a complete idiot for leaving the party without thanking the host."

	THINKING ERROR	DESCRIPTION	EXAMPLE
11	Magnifying or Minimizing	You make a big deal out of something small or minimize positive events in your life.	"I messed up my toast and ruined my sister's wedding rehearsal dinner." "I got a good review on my presentation, but I'm still bad at public speaking."
12	Catastrophizing	You assume things are worse than they are or expect the worst outcome.	"I shouldn't have made that mistake at work. I'm going to get fired."
13	Feelings as Facts	You believe that if you feel something, it must be true.	"I *feel* like I'm out of place, so I must be."
14	Idealistic Comparing	You compare yourself unreasonably and unfavorably to others. You only see the positive aspects in others and measure your life against theirs instead of focusing on your own path.	"I will never be like John. I should be able to have 500 friends on Facebook like he does."

The chart of thinking errors is based on the work of Beck, Burns, and others, including *Feeling Good: The New Mood Therapy*, by Burns, and *Cognitive Therapy and the Emotional Disorders*, by Beck, and adapted to include examples related to social anxiety.

In Real Life: Cara's Speech

My client Cara has been deathly afraid of public speaking, and this dread has impacted her since college, when she had to participate in an on-stage debate. She prepared well, but her heart felt like it was going to jump out of her chest when it was her turn to speak. Cara started shaking, and her mind went blank. She was so short of breath she couldn't speak, and the paralysis forced her to simply say, "I'm sorry, I can't do this" as she stepped off the stage. Cara was terrified this would happen again, and for decades afterward she did everything in her power to avoid public speaking. Then she was assigned to speak on a panel for her job. After Cara considered every possible way out of the situation ("I'll fake being sick . . . I'll miss my plane . . . I'll quit."), she realized it was time to face her fears. Cara sought my help, and we worked on employing one of CBT's core techniques.

Strategy: Cognitive Restructuring

As Cara's negative thoughts escalated in the weeks leading up to the panel ("I'll have a panic attack when the moderator asks me a question . . . I'll forget what I want to say . . . Everyone will see how nervous I am and think there's something wrong with me."), she worked consciously and continuously to restructure her thinking using a thought log. For every negative thought, she used a variety of methods to challenge its validity. She took time to think about the ways she could handle her most-feared scenarios. She generated positive thoughts, the type she would say to a friend in her same situation. Cara got better at talking back to her negative thoughts and not automatically believing them. "Everyone gets nervous sometimes . . . You've prepared well, and even if you forget something, you can look at your note cards."

The Outcome

It worked! Despite the fluttery feelings in her stomach beforehand, Cara was able to immediately defeat all of the negative thoughts that entered her mind with the positive, realistic self-talk she had been developing for weeks. The panel was an incredible success, and the adrenaline rush and newfound confidence motivated Cara to pursue more public speaking opportunities, even if she continues to feel nervous about it.

CBT in Action

When untwisting your socially anxious thinking, you'll probably come to realize that some of the things you are certain will happen are actually not very likely to happen, nor are they based on real evidence. You might also conclude that if the worst-case scenario were to occur, it wouldn't be the end of the world and you would find ways to handle it.

It requires attention to uncover the disempowering stories you are telling yourself that are causing your suffering. It takes practice to identify the distortions in your thinking and come up with positive, believable replacement thoughts.

You can use as many CBT methods as needed until you're able to drastically reduce or eliminate the belief in your negative thoughts. If one method doesn't work, just move on to another one. Remember, you feel the way you think. The moment you no longer believe a negative thought, you'll no longer experience the anxiety, embarrassment, self-consciousness, or other negative feeling that it was causing. As you'll see from the real-life examples in this book, you can use CBT methods to challenge and change negative thoughts in the moment you're anxious, or use them to prepare for an upcoming event that's giving you stress, or engage in them afterward.

As you generate positive, rational thoughts with these exercises, you can write them down in a journal, a notebook, or your phone so you can remind yourself when the negative thoughts come up again. Negative thoughts do keep coming up—that's just human nature—but we can develop ways to take the charge out of them so they don't derail us when they arise.

Here are four important CBT strategies you can use to defeat the negative thoughts that are robbing you of ease and self-esteem. You'll know you've come up with an accurate positive thought that works (and not just positive thinking!) when it immediately reduces your anxiety when you repeat it to yourself. Most importantly, it will drastically reduce or eliminate your belief in your original negative thought. Using these strategies will help you develop a habit of talking back to negative thoughts instead of automatically believing them.

THOUGHT LOG

You can't change your thoughts if you don't know what they are. A thought log, first introduced by CBT founder Dr. Beck, is a tool that helps you figure out and record what you were telling yourself about yourself, the situation, other people, or the future during an anxiety-provoking situation. With your thoughts captured on paper, you can examine, evaluate, and modify them. This was the technique used by Maya in the example on page 27, so review that story for a real-life illustration of how it works.

How to do it:

1. **Record the date and time.**
2. **Describe the situation.** Think of a specific moment that triggered your social anxiety.
3. **Record your emotions.** Write down three feelings you had.
4. **Record your negative thoughts.** Write down one negative thought you had during the situation. What were you telling yourself about yourself, the situation, or the future?
5. **Analyze your distorted thinking.** Review the thinking errors chart on page 28. Choose two possible thinking errors that may be altering your view of reality.
6. **Replace with positive thoughts.** Write down positive things to tell yourself that could defeat the negative thoughts. You can accept some truth in the thought, defend why it's not true, or both.

Thought Log Example:

DATE/TIME	JUNE 10 9 A.M.	JUNE 13 2 P.M.	JUNE 16 5 P.M.	JUNE 20 9 P.M.
SITUATION	Meeting at work	Ordering coffee	Leaving work and ran into coworkers on the way out	Scrolling through Facebook
FEELINGS	Embarrassed Self-conscious Incompetent	Ashamed Defective Worthless	Foolish Inferior Self-conscious	Resentful Discouraged Pessimistic
NEGATIVE THOUGHTS	Everyone thought I was incompetent in the meeting.	I shouldn't be so self-conscious.	My coworkers don't actually want to talk to me. They were just being polite.	I'll never have a circle of friends.
THINKING ERRORS	Mind-reading Magnifying	"Shoulding" Personalizing	Ignoring the Good Mind-Reading	Over-generalizing Idealistic Comparing
REPLACEMENT POSITIVE THOUGHTS	I can't read their minds. Not having a perfect answer can happen to anyone. No one can know everything at all times.	Everyone feels self-conscious sometimes. It shows I care about feeling connected to people. I'm learning new ways to feel more at ease. It takes time.	I don't have evidence they don't want to talk to me. They aren't obliged to ask me about my weekend plans.	I can't predict the future. I'm working on getting to know more people and it takes time. I'll have more opportunities to make connections.

DATE/TIME	SITUATION	FEELINGS	NEGATIVE THOUGHTS	THINKING ERRORS	REPLACEMENT POSITIVE THOUGHTS
JUNE 21 3 P.M.	On my way to a friend's barbecue	Nervous Worried Anxious	I won't know what to say, I'll make a fool of myself, and they won't invite me to another event.	Negative Labeling Catastrophizing	I don't have a crystal ball. Feeling nervous is natural. If I don't know what to say, I can just listen or take a walk and come back.
JUNE 23 6 P.M.	Buying groceries	Humiliated Worried Self-conscious	Everyone in the store saw the stain on my shirt and felt pity for me.	Spotlight Effect Mind-Reading	I don't have proof they felt pity. Some may have noticed, and some may not have. Even if they did, it's not the end of the world.
JUNE 24 10 P.M.	Sitting on the sofa after a dinner date	Despairing Bad Defective	I ruined the date by not asking interesting enough questions.	Filtering Ignoring the Good	It's natural to feel nervous when you're meeting someone new. I can try again on the next date.

DATE/TIME	JUNE 26 8 P.M.	JUNE 30 8 A.M.	JUNE 30 5 P.M.
SITUATION	In the car, returning from a friend's house	In my office before my work presentation	I was invited to happy hour by a few coworkers
FEELINGS	Upset Frustrated Defeated	Panicky Worried Pessimistic	Foolish Discouraged Worried
NEGATIVE THOUGHTS	My friend's husband doesn't like me. He's always judging me.	I'll make a mistake and my coworkers will think less of me.	I won't have anything interesting to add to the conversation.
THINKING ERRORS	Mind-Reading Feelings as Facts	Fortune-Telling Mind-Reading	Filtering Ignoring the Good
REPLACEMENT POSITIVE THOUGHTS	Just because I feel like he's judging me doesn't mean he actually is. His opinion doesn't determine my worth as a human being. Not everyone in life will like me, and this is normal.	I can't know that now. Making mistakes makes me human. If coworkers judge me for this, it's their issue.	I don't have to be interesting to be worthwhile. Listening and being interested is just as valuable as speaking in a conversation.

DECATASTROPHIZING

Catastrophizing is a thinking error, or cognitive distortion, that was first coined by psychologist Albert Ellis in the 1960s. When you're catastrophizing, you predict the worst possible outcome and underestimate your ability to cope.

Decatastrophizing involves confronting your worst-case scenario, challenging the validity of your thinking, and shifting the focus to how you could handle things if the worst outcome occurred. By learning to question your catastrophic thinking, you discover that the worst is not likely to happen, and you are more capable of coping than you assumed.

How to do it:

1. Think of an anxiety-provoking social situation and write down one thing you are worried will happen. Be as specific as possible.

   Examples: *She'll turn me down. I'll freeze up and embarrass myself. My face will blush.*

2. How likely is this event to happen? Are you basing your prediction on feelings or facts?
3. What's the worst scenario? What's the best scenario? Would it be as awful as you think?
4. If the worst did happen, how could you cope? Write down what actions you could take to handle the situation.
5. Will you care about this one week from now? In one month? In one year? In five years?
6. Write a reassuring, positive replacement thought you can tell yourself that defeats your original worry. What would a friend say?

THOUGHTS ON TRIAL

Also called Examine the Evidence, this is a CBT technique for examining and reality-testing irrational thoughts. The exercise can help you question information that's supporting a negative thought by challenging you to review the evidence for and against the belief, including evidence you may have ignored.

How to do it:

1. **Set the scene.** Describe a social situation in the last six months when your anxiety was high. Where were you? Who were you with? What were you doing?
   Example: *I had to make a 15-minute presentation during a work meeting attended by my boss and 10 colleagues. I showed 12 PowerPoint slides, projected on a screen from my laptop, and I explained each slide while seated at the conference table.*

2. **Identify the thought.** Write down one negative thought that went through your mind at the time.
   Example: *Everyone can tell how nervous I am and it's ruining my presentation.*

3. **Put the Thought on Trial.**
   Column 1. Defense.

   What evidence supports this thought being true? Only include the facts.
   Example: *My voice was shaky. People asked me to speak up. I had to pause halfway through to correct an error on one of the slides.*

   Column 2. Prosecution.

   What evidence supports this thought being false? Only include the facts.
   Example: *When I finished, several people, including my supervisor, spoke up and said they were impressed by my work. Two people said they had a better understanding of the subject because of how I'd explained it.*

4. **Render a verdict.** In light of the evidence presented by the defense and prosecution, which side won? Do you still believe your negative thought? If not, what could you tell yourself instead that is accurate and balanced?
   Example: *My negative thought was not accurate. My presentation may not have been as smooth as I would have liked, and I need to practice so I can speak more comfortably. But I got positive feedback, and the group seemed to understand and appreciate what I was telling them.*

WHAT WOULD I SAY TO A FRIEND?

You're a total idiot! You're going to embarrass yourself! You completely blew it! We would never say these things to a dear friend who was feeling nervous or self-conscious, yet that's what our inner critics sound like inside our heads. When a negative feeling or thought comes up, you can shift your perspective by asking yourself what you might say to a close friend or family member in the same situation. Thinking in this way encourages us to practice speaking to ourselves in kinder, gentler, more compassionate ways.

How to do it:

1. Think of a recent social or performance situation that didn't go as well as you had hoped. Write down three negative thoughts you were telling yourself.

   Example: *I looked like an idiot. I froze up and put him off. I screwed up my solo.*

2. Now imagine a close friend was telling you these things aloud. What is something truthful, compassionate, and encouraging you could tell them?
3. Now think of a social and performance situation that is coming up that you are certain will go badly. Write down three negative thoughts you are telling yourself.

   Example: *It will be obvious what a loser I am. She'll see I have no personality. He'll think I'm ugly. They'll notice how bad my social skills are. I won't be able to express myself.*

4. Now imagine, again, a close friend was telling you these things aloud. What is something truthful, compassionate, and encouraging you could tell them?
5. Now think of some of the self-critical things you say to yourself about having social anxiety. Write down three negative thoughts you tell yourself.

   Example: *I shouldn't be so awkward. I should be more skilled at talking to people. There's something wrong with me.*

6. Now imagine, again, a close friend was telling you these things aloud. What is something truthful, compassionate, and encouraging you could tell them?

In Real Life: Rogerio's Date

For years, Rogerio has been longing to fall in love. A couple of days before a date, he usually feels pretty optimistic. But when the day arrives, his anxiety kicks into gear and he's lost in "what if" thinking. Thoughts of making eye contact, eating in front of someone, and having to make small talk make his stomach feel queasy and his heart race. Sometimes he cancels at the last minute—he has standard excuses that he uses—but afterward he beats himself up for not following through. Other times, he manages to get himself out the door, but the anxiety persists throughout the date and for hours afterward.

Strategy: Cognitive Restructuring

Rogerio uses cognitive restructuring practice to help dating become easier. A few hours before a date, he creates a thought log, writing down all of the worst things he's predicting will happen, such as "I'll appear foolish," "I won't know what to say," or "My date won't want to see me again." He's gotten faster at noticing when his mind is catastrophizing, fortune-telling, or mind-reading, and has replacement thoughts ready to defeat these cognitive distortions. Rogerio tells himself, "It's okay to be nervous," "I don't have to be perfect," and "I don't have a crystal ball and can't know now if the date will fail."

The Outcome

A few months ago, Rogerio finally met a man he could imagine spending the rest of his life with. In the car on his way to meet Grant for their first dinner date, Rogerio was apprehensive and on edge. This time, he was able to talk to himself positively and rationally, challenging his predictions of how badly he thought the date would go. During dinner, he felt a little more relaxed, and enjoyed more eye contact than usual. Each date with Grant has gotten easier, and they are even planning a weekend trip away, something Rogerio has been waiting to experience for a long time.

Behavioral Experiments

As we've explored in this chapter, recognizing your negative thinking patterns, identifying what thinking errors you're using to support them, then replacing them with realistic and balanced thoughts can immediately shift your outlook. Next, we'll look at another approach for unraveling negative thoughts. You're about to test the accuracy of your self-defeating beliefs by becoming the scientific director of your very own behavioral experiment.

In cognitive behavioral therapy, a behavioral experiment (BE) is set up to test unhelpful beliefs through experiential learning—"learning by doing." Behavioral experiments put us in a real-world situation, where our actions and observations will yield new information relating to the beliefs we're testing. BEs can generate a more positive, rational, and accurate view of a situation that triggers our social anxiety. This can even happen with just one experiment. Before we discuss the details in planning and conducting your behavioral experiment, let's look at an example.

Moshe, a former client, came to therapy to overcome his fear of social gatherings, which was holding him back from the connection and community that he wanted. He had been turning down invitations to parties, weddings, graduations, and most recently his niece's bat mitzvah. His sister was disappointed, and he felt guilty and ashamed.

After several CBT sessions, Moshe was able to identify some of the negative thoughts that were causing him the most anxiety at parties, including "I'm out of place" and "Everybody thinks I'm boring and doesn't want to talk to me." Moshe was able to identify some of the cognitive distortions that reinforced his negative thinking, like feelings as facts and mind-reading. He came up with some replacement thoughts: "Just because I feel out of place doesn't mean I *am* out of place . . . I can't read people's minds or please everyone." However, he was still convinced that his negative thoughts were somewhat true. We decided that a behavioral experiment would be an effective way to put this to the test.

For his experiment, Moshe formed this hypothesis: "If I go to a party, people will notice how out of place I am and won't want to talk with me." He then rated how certain he was of this prediction: 70 percent. Together, we discussed what kind of BE would enable him to gather evidence. Moshe decided to conduct his experiment at his niece's bat mitzvah the next month. His plan was to attend the service and stay at the reception that followed for no longer than 45 minutes. During the reception, he would start a conversation with four people. He would observe if they continued the conversation with him for more than two minutes before moving on to someone else. Moshe predicted that only one person would want to talk to him.

Moshe came to his therapy session following the bat mitzvah with a big smile on his face. It was clear the behavioral experiment had made a huge impact on him. Moshe was surprised to discover that four people talked with him for more than 10 minutes. He even talked with two more people on the way to his car. "I now understand on a gut level that I'm not as out of place as I thought was true," he told me. He was equally surprised by how well the conversations went and how enjoyable it was to connect with people, even for just 10 minutes. And his sister had shared with him how grateful she was to have him there, which lifted his spirits.

Adult learning theories suggest that both experience and reflection create the most effective learning. With some guidance, Moshe was able to reflect on his discovery. Taking into consideration what the experience taught him, he re-rated his belief in his original prediction at zero percent. Moshe didn't believe his negative thought, "If I go to a party, people will notice how out of place I am and won't want to talk with me," any longer. He composed a more realistic replacement thought: "Even if I *feel* like I'm out of place, that doesn't mean others think so. I can create a connection with other people." Moshe decided to write this new thought on a sticky note and keep it posted on his fridge as a reminder when tempted to turn down an invitation in the future.

BEHAVIORAL EXPERIMENT

You might have an idea of what will go wrong in certain social situations. But are you sure? One way of testing that belief is to conduct an experiment. Like a scientist in a lab, you'll form a hypothesis, test it with an experiment, then examine the results.

How to do it:

1. Start by writing down a social situation that you predict will turn out badly.
2. Narrow down a particular assumption to test. What is the belief you are testing? What do you expect will happen in this situation? Be as specific as possible. Try using an if/then statement to be precise.

   Examples:
   If I eat in front of others at a restaurant, they'll think I'm gross.
   If I go on a date, I'll have nothing interesting to say.
   If I speak up in class, my classmates will laugh at me.
   Even if I try to talk to people at a party, I'll embarrass myself and they won't talk to me for more than a minute before moving on to someone else.
3. How strongly do you believe that this prediction is accurate? Rate your confidence from zero to 100 percent.
4. It's a good idea to choose a setting that causes low anxiety and keep the experiment short. With that in mind, consider the following questions to determine a situation that will test your prediction:

   Where do you need to go, and what do you need to do or observe?

   How long will it last?

   Do you need someone to go with you?

   How many times might you need to repeat the experiment?

   What data will you capture, and how will you record it?

   Example: *I need to be in a social situation with people I haven't met before so I can see if strangers will talk to me for more than a minute. To test this, I'll go to my friend's graduation party by myself. I will stay there for at least 30 minutes and try to start a conversation with at least three people I don't know. I'll use the stopwatch function on my phone to observe how long each person talks to me, and after each conversation I'll jot down how long it lasted in my phone's notes app.*

5. Carry out the experiment and record your observations.
6. Review your findings and answer the following questions:

   Was your prediction accurate?

   How strongly, from zero to 100 percent, do you believe your original prediction now?

   What did you learn from the experiment?

   Example: *I ended up talking to six different strangers, and in four cases we spoke for more than 15 minutes. One person excused himself after about a minute to use the restroom. My prediction was not accurate, and I no longer believe it. I learned that I can be comfortable at a party for longer than I expected. I planned to stay for 30 minutes but lost track of time and ended up staying for almost two hours!*

   Given the evidence from the experiment, what is something positive and realistic you can now tell yourself?

   Example: *I can't know if people are willing to talk with me unless I try.*

CHAPTER THREE

Get Out There

Fear keeps us playing small. As long as we're protecting ourselves from the outside and trying to avoid social situations, we'll continue to live in fear. In this chapter, you'll have a chance to play big. This chapter is about exposure therapy, one of the most powerful ways to face your fears and build your confidence in the process. You'll have opportunities to create your own hierarchy of exposure to social fears and try out different types of exposure therapy exercises to move through your own fears.

We'll look at the history of exposure therapy and see how to apply its principles to shift your relationship to fear. We'll investigate how leaning into discomfort is exactly the solution for feeling comfortable in your own skin. We'll examine the variations of exposure therapy, from real-life practice to virtual reality environments. We'll explore the different ways you can pace exposure treatment, from gradual exposure to facing your biggest fear all at once.

We all have the natural ability to break old habits at any moment. We often think we need confidence *before* we can do something difficult. In reality, it's when we do hard things that we grow our confidence. We already have everything we need to be courageous.

Exposure Therapy: A Primer

As we discussed in the last chapter, escaping situations you're afraid of can bring a short respite, but in the long run, your anxiety worsens. Exposure therapy is a method of interrupting this pattern of avoidance. The concept is a simple one: In a safe environment, you're gradually exposed to the kind of situation that causes you distress until your fear diminishes or completely goes away. It's like climbing into a hot tub. At first, the heat of the water is almost intolerable. But if you stay in the water even just a little longer, your body's temperature adjusts. After a little while, you start to feel comfortable, and even enjoy the soak and not want to get out!

This therapeutic approach goes back to the work of psychologists from the early 1900s, like Mary Cover Jones, who began to understand that behavioral responses like fear could be "unlearned." By the 1950s, a treatment called systematic desensitization was invented by South African psychiatrist Joseph Wolpe to deliberately expose patients to their fears to reduce sensitivity. Since then, exposure therapy has been developed as one type of cognitive behavioral therapy. Exposure therapy's been scientifically demonstrated to be a highly effective treatment for a whole range of problems, including social anxiety, phobias, panic disorder, obsessive compulsive disorder (OCD), and post-traumatic stress disorder (PTSD).

For social anxiety, exposure therapy can help people overcome fears of specific social and performance situations, such as meeting new people or giving a talk. It's typically conducted with the guidance of a therapist who's trained and experienced in exposure therapy as part of a cognitive behavioral therapy program. Most people start with situations that are minimally unpleasant, then slowly and methodically face their fears until they are ready to face what is most frightening. In this chapter, you'll learn some safe and simple exposure therapy techniques that you can use on your own, though as always, a trained therapist can offer a more advanced treatment program.

When you're using exposure therapy on your own, accountability is what's most important. For example, if your anxiety levels rise during an exercise, you want to make sure you don't quit before you've given yourself the chance to break through it. Consider having someone in your life—a friend or family member—assist you with being accountable and join you when it's helpful. For example, you could have a friend come with you during a social mishap exercise (an exercise we'll describe later in this chapter). If you're concerned you may have a severe reaction to an exposure exercise or have a significant health issue like a heart condition, it's best to consult a doctor or mental health professional.

In the beginning, when you're confronting what scares you, you might feel a little uncomfortable or uneasy. But remaining in an anxiety-provoking situation enables you to experience—firsthand—that what you've been dreading would happen doesn't actually happen, that the situation isn't as bad as you imagined, and that you can effectively handle it (especially with the cognitive restructuring strategies from chapter 2 to help you). By being brave enough to expose yourself to the situations you fear, you can build confidence and tenacity as you become habituated to the process of facing your fears.

It's hard not to feel powerless when fear stops us from living the kind of life we want for ourselves. But during exposure therapy, you get to experience what it feels like to take back control of your reactions. As your confidence grows, facing your fears becomes empowering and even enjoyable. Countless clients of mine have begun a therapy session by sharing how exhilarated and free they felt, for the first time in their lives, after completing an exposure therapy exercise. For example, one client went to her favorite art store and asked a cashier to use an expired coupon, something that in the past would have made her cringe with embarrassment. She told me she was utterly surprised by how happily and willingly the cashier accepted her coupon. Her feelings of excitement lasted for the rest of the weekend.

And the effects of exposure therapy can go beyond easing social anxiety. Even after only brief practice with exposure exercises, clients tell me they feel less anxiety in general in their day-to-day lives. Their brains and bodies react completely differently to previously feared situations, much of which stemmed from fear of the unknown.

Here are three major benefits you can expect from using exposure therapy to conquer your social anxiety:

The feared becomes familiar. Repeated exposure to something you're afraid of causes *habituation*, which means you're so accustomed to it that you stop responding or even noticing it. You're so used to it that the situation is no longer fearful. Whatever nervousness remains is minor and likely to fade quickly once you get started.

The fear goes away. Sometimes exposure therapy leads to *extinction* of your fear response: the thing that scared you no longer scares you. The majority of my clients completely extinguish their social anxiety after sufficient exposure practice. For example, they've practiced rejection exercises out in the world repeatedly, to the point that they not only don't experience anxiety about being rejected, they're unfazed by negative reactions or rejection from others.

You believe in yourself. In my practice, I've never seen anything grow someone's confidence as rapidly as practicing exposure exercises. Over and over again, clients have told me that successfully participating in exposure therapy showed them they were stronger than their fears and more capable of handling challenging situations than they could have possibly predicted. They experience what's called *self-efficacy*, the belief in your own ability to learn and master skills to cope effectively with challenging situations. If you face your fears and tolerate them without avoiding, you begin to realize you're more resilient than you assumed. And that, in turn, makes you more willing to face fears in other contexts.

We can't avoid the hard stuff if we want to reach our goal of freeing ourselves from social anxiety. But every time you face what you're afraid of, you'll gain momentum and feel empowered for the next challenge. Exposure therapy allows you to deeply experience that you are capable of managing your fear and anxiety. Imagine no longer having to rearrange your life to avoid facing your fears. There is freedom on the other side.

Variations of Exposure

There are several ways to try exposure therapy. Here's a look at the variations that will help you think about which approach is best for you, whether you're trying it on your own or working with a therapist.

In Vivo Exposure

The most common form of exposure is *in vivo* exposure, in which you experience a feared situation in a real-life setting. (The name is from Latin, meaning "in the living.") For someone who has social anxiety, this means getting out in the world and doing what you're afraid of, from starting a conversation with a stranger to giving a speech. The point is to directly experience your anxiety-provoking situation in the real world, to reduce your fear.

During in vivo exposure, you learn to stay in the situation you dread until your fear subsides. When that happens, your belief in your own abilities will dramatically change.

Research shows that in vivo is the type of exposure therapy that works best for social anxiety. Other methods are also effective, but they work even better when used with an in vivo approach.

Imaginal Exposure

An alternative to real-life contact with a social anxiety trigger, *imaginal* exposure therapy involves vividly imagining the feared situation. As with in vivo, the idea is to evoke your fear and anxiety and stay with it until it dissipates. The aim is to grow more comfortable with the feared situation so your fear response will be reduced when you confront the same circumstances in real life. Imaginal exposure can be a great way to start with exposure therapy, especially if the idea of trying things out in the real world seems overwhelming.

A session of imaginal exposure is a mental exercise in which you imagine the event or situation that you fear with as much detail as you can muster and visualize yourself facing it. The imagined scenario can be experienced like a movie, and the exercise is often guided by a therapist, who may use a script. You're encouraged to close your eyes and vividly imagine scenes that evoke anxiety and discomfort, concentrating on sensory elements like sounds, tactile sensations, and smells to make everything seem more real. The script might be recorded so you can listen to it on your own.

When you're doing imaginal exposure on your own, you can sit quietly, close your eyes, and use your imagination to put yourself in a situation that invokes your fear. You can also record your own script, then play it back to yourself with eyes closed. Talk yourself through the scene, encouraging yourself to imagine each detail.

Virtual Reality Exposure

Virtual reality exposure therapy (VRET) is a high-tech method that presents feared situations through computer-generated scenes using a virtual reality headset. During a VRET session, you wear a head-mounted display system with binocular screens and are immersed in a virtual environment that exposes you to anxiety-provoking stimuli through computer-generated displays. Like a pilot in a flight stimulator learning how to fly a plane without crashing, you can learn to confront your feared social or performance situation in a safe environment that might be harder to do in real life. VRET is an expensive technology and not as widely available as other types of exposure therapy.

Interoceptive Exposure

This approach focuses more on the physiological aspects of fear and anxiety than the situation that provokes them. Interoceptive exposure allows you to experience the bodily symptoms of anxiety in a controlled way so you aren't thrown off when they happen in a real-life situation. In this therapy, you will deliberately bring on physical sensations like sweating or a fast heartbeat so you can confront your fear of experiencing these reactions. For example, you may be instructed to run in place or do jumping jacks in order to make your heart speed up, with the aim of learning that this sensation is not actually dangerous.

This kind of therapy is especially helpful for people who have what's called anxiety sensitivity (AS), a tendency to misinterpret the physical sensations that accompany anxiety as something scary and dangerous. If you're prone to AS, you might think you're going to have a heart attack or suffocate. You might dread sweating or trembling in front of others. By re-creating the bodily sensations that can happen during periods of intense anxiety, you'll build your tolerance for them.

In Real Life: Celeste's Introduction

At the start of therapy, Celeste told me that introducing herself to others was one of her biggest challenges. She shared that this used to be easier when she was in her 20s. Since becoming a stay-at-home mother, however, she'd become more isolated and hadn't been able to introduce herself to any parents at the playground. Despite feeling desperate for adult interaction, Celeste was out of practice, and she dreaded being around people she didn't know.

When her daughter started preschool, Celeste imagined she would overcome her anxiety once and for all, and finally start making new friends. Instead, her anxiety increased. She was nervous at drop-off and pick-up every day. She became even lonelier, longing to connect with other moms for playdates. With the school year half over, she was discouraged that she still hadn't worked up the courage to introduce herself to anyone.

Strategy: Imaginal Exposure

Determined to get past feeling awkward, Celeste created a goal for herself to exchange phone numbers with one mom from the preschool. Celeste started by imagining introducing herself to someone after school. With eyes closed, Celeste visualized the scene in vivid detail, from what she would be wearing, to the sound of the leaves rustling in the trees in the front yard of the preschool. She developed a script of what she would say. She practiced imaginal exposure regularly until she felt ready to do it in real life.

The Outcome

At pick-up one afternoon, after their kids had been playing together in the yard for a few minutes, Celeste finally got up the nerve to practice her script with another mom at the preschool. She said, "Hi, I'm Celeste. I don't think I've officially introduced myself." The other mom smiled and remarked, "It's funny how us parents know all of the kids' names, but we forget to mention our own!" Celeste was relieved and encouraged by the friendly response. After a few days of casual chatting while their kids played, Celeste finally suggested they have a playdate, and they exchanged phone numbers. The playdate went so well that Celeste and her daughter are now enjoying weekly meetups at their local park.

15 WAYS TO PLAN A SOCIAL MISHAP

Ready to plan an in vivo session of your own? A good starting technique is to try some social mishap exercises. Like the name suggests, in these exercises you'll challenge yourself to make a mistake in a social situation. You'll get rejected on purpose or break a social rule—in a minor way, of course. The purpose is to lessen your fears and challenge your assumptions about people's reactions to you.

The exercises may seem trivial, but they provide some of the greatest opportunities to completely diminish your social anxiety. Believe me, doing them will grow your courage and confidence immediately. I've had many clients tell me how, after completing a social mishap, that they were surprised by how generous, kind, and forgiving people turn out to be. Or how rarely people were bothered by their "odd" behavior. Doing these exercises can actually be thrilling and exhilarating! For bonus points, I recommend watching "What I Learned from 100 Days of Rejection," a TED Talk by Jia Jiang. He shares how doing social mishap exercises for 100 days straight enabled him to overcome his fear of rejection. He even developed an app based on the concept, called DareMe.

How to do it:

Here are 15 easy social mishap exercises to try. You can do these in order, or start with the ones that seem most doable, or choose at random. You can break an exercise into smaller steps if you like. (For exercise 1, you might start by making eye contact the first time, then smile at someone the next time, then finally say "hello" to a third stranger). You can skip around the list or repeat one exercise until your fear subsides, then move on to the next.

1. Walk down the street and say hello to one stranger.
2. Ask a stranger to borrow one dollar.
3. Offer to give one dollar to a stranger.
4. Ask someone, "Is it your birthday today?"
5. Ask to buy a stranger a cup of coffee.
6. Ask someone to buy you a cup of coffee.
7. At a store, ask for a discount or to use an expired coupon.
8. Ask for directions to a place you're standing right in front of.
9. Walk down the street carrying an open umbrella on a sunny day.
10. Sit down in the middle of a busy sidewalk and read a book for one minute.
11. Ask someone if you can skip ahead in line.
12. Ask someone if you can borrow the book they're reading.
13. Ask someone if they like your shoes.
14. Run and get one more thing when you're about to check out at the grocery store.
15. Ask someone if they like your sunglasses or eyeglasses.

Before you start, write down some predictions of what you think will happen when carrying out the exercises. What will other people do? What will they say to you?

Afterward, write down what actually happened. What were others' responses? What were you surprised by? How did you feel? What did you discover?

INTEROCEPTIVE EXPOSURE EXERCISES

For some people, the physical sensations of social anxiety are as upsetting as their fear and nervousness. Interoceptive exposure exercises produce physical sensations that mimic the body's fight-or-flight response, with the goal of desensitizing you to those physical feelings. The more you experience them in a controlled setting, the more you will get used to your body's reactions, and the fear will gradually subside.

Here are three simple exercises that provide interoceptive exposure; each replicates a different set of symptoms that you might experience when your social anxiety is active. When you've become familiar with each exercise, move on to the bonus exercise. Please note that if you have any medical conditions, you should check with your doctor before attempting any of these exercises, especially if you have epilepsy or seizures, have a heart condition, have low blood pressure, have physical injuries, have asthma or lung issues, or are pregnant.

How to do it:

The exercises are given in no particular order. Try all of them, starting with the exercises that seem the least difficult. Then choose an anxiety sensation (e.g., shortness of breath or rapid heartbeat) that you experience in your life and practice the relevant exercise daily. These exercises work best when you bring a willingness to experience the sensations instead of resisting or struggling against them. After each exercise, be sure to write some notes on your experience and rate your anxiety about the sensations on a scale from zero to 10.

EXERCISE 1:

Helpful if your anxiety produces shortness of breath, racing heart, and sweating.

1. Run in place for one minute.
2. Stand still and notice your body's sensations (don't drink water yet or go outside to cool off).
3. Notice your heart pounding.

4. Notice your breath.
5. Notice your body sweating.
   - Repeat three more times.
   - Write down the outcome. What sensations are you experiencing? What did you learn?

Bonus exercise: Run up and down a flight of stairs for one minute.

EXERCISE 2:

Helpful for tension, shaking, and tightness in the stomach.

1. Tense all the muscles in your body, holding them tense for one minute.
2. Notice any shaking.
3. Notice any tightening.
4. Notice where you feel tension in your body.
   - Repeat three more times.
   - Write down the outcome. What sensations did you experience while you were tensing your muscles? How do you feel now? What did you learn?

Bonus exercise: Tense your muscles for more than one minute.

EXERCISE 3:

Helpful for tolerating shortness of breath, feeling like you can't breathe, or feeling that you might suffocate.

1. Take normal breaths through a small drinking straw for one minute.
2. Don't take any deep breaths afterward.
3. Notice your breathing. What does it feel like?
4. Notice any tightness in your throat or neck.
   - Practice three more times.
   - Write down the outcome. What sensations are you experiencing? What did you learn?

Bonus exercise: Hold your nose closed while you breathe through the straw.

Setting the Pace

While exposure is one of the most powerful methods for overcoming fears, using this type of therapy effectively can take some planning and patience. Exposure therapy, whether in vivo or imaginal, can be paced to match your comfort level and the intensity of your anxiety.

Graded Exposure

One of the most effective techniques, and also the most common form of exposure therapy, graded exposure means slowly and progressively exposing yourself to your fears in order to reduce your sensitivity gradually and at a comfortable pace. The first step in this approach is to identify what specifically makes you socially anxious. With this information, you create an "exposure hierarchy," a ranking of feared events or situations in order of severity. Typically, scenarios are ranked on a 1 to 10 scale, with 1 causing mild distress and 10 causing extreme anxiety.

For graded exposure, you'll start by confronting a mildly anxiety-provoking situation, something that causes some discomfort but that you can handle (making chitchat with a coworker, for example). From there, you slowly work your way up the hierarchy, with the experiments increasing in difficulty. Each step of the way, you can use whatever mix of in vivo, imaginal, and interoceptive exposure exercises are most effective. You'll stay with each situation, repeating the exposure until the task becomes easy, and move to a more challenging level when you feel comfortable. The goal is to master your anxiety and fear at each stage.

Systematic Desensitization

Similar to graded exposure, systematic desensitization exposes you to your fears in gradual steps, following an exposure hierarchy that you create. But with this approach, the exposure is paired with relaxation exercises to make the steps more manageable. The fear response you would normally have is replaced with a relaxation response, a state that's incompatible with fear and anxiety. You learn breathing exercises and muscle relaxation techniques. You might also use visualization, or guided imagery, in which you imagine a relaxing scene in your mind and focus on sensory details, like the colors, sounds, or smells around you.

Flooding

Flooding is the most extreme and intimidating form of exposure therapy. It also uses the exposure hierarchy, but instead of starting at the bottom of your list, you begin with the most feared situation. At first, flooding produces extreme anxiety, maybe even panic. But anxiety levels can't be sustained indefinitely. Once the panic subsides, you're able to realize you haven't experienced any harm, and the fear is extinguished.

That probably sounds intense, and it can be! But the technique can be very useful if you're up for it and motivated to get over your fears as quickly as possible. Flooding produces results quicker than graded exposure. But it's not an appropriate treatment for all people and all situations in which they experience social anxiety.

In Real Life: Alana's Networking Event

Since she was young, Alana had an ongoing intense fear of interacting with strangers. In college, determined to find her first job, she found the courage to attend a career fair. As she approached her favorite company, résumé in hand, she was already trembling and sweating. When she tried to speak with the recruiter, her mind went blank. She was mortified as she stood in awkward silence and left abruptly. Alana felt like a failure and spent the rest of the day blaming herself for being socially awkward and ruining her own chances for her dream job.

After college, Alana was able to successfully avoid all networking events until she reached a dead-end at her job. Since most of her friends were getting jobs through these events, Alana would register for one, dread it for weeks, and then back out at the last minute.

Strategy: Graded Exposure

When Alana came to see me, she was discouraged about missing out on job opportunities and was eager try a new approach. She was willing to try gradual exposure therapy, and together we developed an *exposure hierarchy*, a ranked list of fearful situations involving talking to strangers. Gradually exposing herself to her fears (a graded, in vivo approach), Alana worked her way through each level of her hierarchy, starting with the easiest, saying “hi” to a stranger, moving on to conversing with the barista at her local coffee shop, and so on. Once she was comfortable with each level of social contact, she would move on to the next.

The Outcome

The highest-ranking item on her exposure hierarchy was attending an upcoming job fair. Alana was determined to stick it out no matter how anxious she felt. After 15 minutes of talking with new people at the job fair, Alana was surprised to discover her discomfort and uneasiness had completely subsided. By intentionally exposing herself to what she had most feared, she got to experience a new level of relief. She felt free to talk to employers from her favorite companies for the rest of the evening. She made a valuable connection that day, which led to a new career. What had once been terrifying was now enjoyable to her, and even fun.

YOUR LADDER OF COURAGE

Graded exposure therapy can grow your confidence as you face increasingly difficult anxiety-provoking situations. As you move forward, you might discover the consequences are not as bad as you had imagined. The first step in using graded exposure to conquer your social anxiety is to create your own hierarchy of exposure (I also call it your ladder of courage). Here's how to create your own hierarchy and start using it to gradually overcome social anxiety. It's natural to feel apprehensive before exposing yourself to a social anxiety trigger, even a mild one. But if you can follow through, the rewards are huge. If you reach a point on your hierarchy that seems too difficult, I recommend you seek the help and guidance of a cognitive behavioral therapist trained in exposure therapy.

How to do it:

1. **Define the problem.** Describe a social or performance situation that makes you feel anxious. If you had to put a name to the kind of social encounter that most triggers your social anxiety, what would it be?

   Examples: *Talking to someone in authority, being the center of attention, performing in front of an audience, speaking in meetings, having a conversation with a stranger.*

2. **List some examples.** Think about the social anxiety trigger you named in step 1. Write down as many specific situations as you can think of when your anxiety comes into play. Think about occurrences that happened in the past and situations that you particularly dread. Consider instances in an ordinary day when your anxiety comes up as well as notable events made more difficult by your anxiety.

   Example: *If your step 1 answer was "Speaking in front of other people," you might list "answering a question in class," "talking to customers at work," "ordering a pizza over the phone," "presenting a research paper," "giving a toast at my sister's wedding."*

3. **Create a ladder.** Now review your list and rank each situation on a scale from 1 (causes the least discomfort) to 10 (causes the most fear possible).

   Rewrite your list, arranging fears in order of their ranking, making sure you have an example for each rung of your ladder. You may want to replace your most-feared situation with something you specifically want to achieve, like Alana's goal of attending a job fair.

   If you feel too overwhelmed to face a situation at once, you can break it down into smaller, more manageable steps.

   Example: *Molly would like to work on her anxiety about attending happy hour with coworkers after work. Her ladder of courage looks like this:*

Anxiety-Inducing Situation	Ranking
• Go to happy hour with coworkers	10
• Go to restaurant/bar with a few coworkers after work	9
• Invite three coworkers to dinner	8
• Invite three coworkers to lunch	7
• Go to lunch with coworkers	6
• Invite one familiar coworker and one acquaintance for coffee	5
• Invite two familiar coworkers for coffee	4
• Invite one familiar coworker for coffee	3
• Ask five coworkers about their weekend plans	2
• Say hi to five coworkers	1

4. Start climbing the ladder. Choose an anxiety-inducing situation that you ranked at 1 or 2. Commit to facing your fears by participating in this situation until your anxiety rating is minimal or extinguished. Once you feel comfortable with one rung, you can move up to the next.

At the start, you'll find that during each situation your anxiety levels will rise, reach a peak, then eventually go down. When your anxiety peaks, this is when you will most likely feel like escaping. It's critical at this point, however, to stay with the fear, even for a few more seconds, to give yourself a chance to see that you can handle it. You'll also find that as you progress along your exposure hierarchy, the idea of moving on to a more difficult task becomes less intimidating. You may even look forward to challenging yourself with something new.

SYSTEMATIC DESENSITIZATION WITH IMAGINAL EXPOSURE

Systematic desensitization is a two-pronged approach to conquering social anxiety: Graded exposure helps you gradually become comfortable in social situations, while relaxation techniques make each step easier. In this exercise, you'll learn a basic relaxation method and then use imaginal exposure to work through your hierarchy of exposure. You can also pair systematic desensitization with in vivo exposure once you know the relaxation technique well enough to employ it in real situations.

How to do it:

1. **Relaxation**
   Relaxation techniques help you activate a calm response in your body, one that's incompatible with anxiety. Essentially, you're switching off the fight-or-flight stress reaction and switching into rest-and-digest mode. Practice this easy breathing exercise to replace anxiety with relaxation.

   4-7-8 Breath

   - Sit comfortably in a chair or lie on the floor.
   - Place your hand on your stomach so you can feel your diaphragm move up and down as you breathe.
   - Inhale slowly through your nose for a count of 4.
   - Hold your breath for a count of 7.
   - Exhale slowly through your mouth for a count of 8.
   - Repeat this breathing cycle for 5 minutes.

2. **Imaginal exposure**
   Review the hierarchy of exposure exercise you completed previously. Continue taking deep breaths as you visualize the least anxiety-inducing situation on your list.

   Let the image in your mind be as vivid as possible. If you were actually experiencing this situation, what would you see around you? Who would be there? Imagine the sights and smells. Who are you with? What time of day is it?

   If you're able to stay completely relaxed while visualizing the first situation on your hierarchy, you're ready to move to the next step.

CHAPTER FOUR

ACT and Commit

Imagine you're going on a trip, and the destination is a place you've wanted to visit for many years. When you arrive at the train station, you see there are two types of trains you can take. One is ready to depart but has strange-looking train cars with hard, uncomfortable seats. The other looks safe and familiar, with air-conditioning, and even a fancy dining car.

You keep waiting for the safe, luxurious train to start moving, while on the other platform one strange, uncomfortable train after another keeps leaving the station. What if that comfortable train won't ever depart? Isn't it better to accept discomfort and get to your destination?

That metaphor, originated by ACT clinician Aidan Hart and appearing in *The Big Book of ACT Metaphors,* illustrates the focus of acceptance and commitment therapy (ACT), the subject of this chapter. If we know where we want to go in life, what are we willing to experience to get there? In this chapter, you'll learn acceptance and mindfulness exercises that, together with action and goal-setting strategies, will help you handle the unpleasant feelings and thoughts of social anxiety so they don't stop you from doing what's important to you and going where you need to go.

Acceptance and Commitment Therapy: A Primer

If acceptance and commitment therapy had a slogan, it would be "Embrace your demons, and follow your heart!" according to Russ Harris, MD, author of *The Happiness Trap*. The main objectives of this therapeutic approach are to *accept* what's out of your control and *commit* to action that improves your life.

Acceptance and commitment therapy (ACT) encourages you to create a full and meaningful life—follow your heart—while accepting that life will always include pain, loss, rejection, and suffering. Trying to control, lessen, or suppress your social anxiety symptoms is not only ineffective but detrimental to your well-being. Instead, ACT teaches you how to accept that the symptoms exist—embracing your demons, or at least letting them come along for the ride. In doing so, you can unhook from those difficult feelings and thoughts so you can pursue your goals without being restrained by them.

Created by psychologist Steven C. Hayes in the 1980s, ACT (the acronym is pronounced as the word "act") is considered part of the "third wave" of cognitive behavioral therapies. The first wave started in the 1950s with Ivan Pavlov's classical and operant conditioning experiments. (You may have heard of Pavlov's famous experiments, in which he found that dogs could be trained to salivate not just in response to food but to a neutral stimulus like a bell.) The second wave emerged in the 1970s, which included the start of cognitive behavioral therapy. While the first two waves of behavior therapy are focused on the *content* of your internal experiences, the third wave focuses more on the *process* of how you relate to your internal experiences.

When you look under the hood of ACT, you find that the engine runs on mindfulness. Being able to stay in contact with the present moment, despite difficult thoughts and feelings, is a key ACT principle known as *psychological flexibility*. Mindfulness, the ability to focus one's attention on the present moment and keep it there, is the method for achieving that flexible state. ACT teaches three important mindfulness skills to get you there: *defusion*, *acceptance*, and *being present*. We'll explore those concepts more deeply in a bit.

Along with practicing mindfulness, ACT encourages you to identify your values, your heart's deepest desires for how you want to be and how you want to treat others. These values serve to motivate and guide you to make the positive changes necessary to have an inspiring existence. For example, if you value being a great friend, ACT empowers you to make time to be with your friends and do thoughtful things for them. If you value health and well-being, ACT helps you prioritize exercise, eating well, and practicing self-care.

ACT techniques can be applied to social anxiety in a variety of ways. Learning to separate from anxious thoughts, instead of getting caught up in them, is an important first step in reducing their power over you. The ability to make room for uncomfortable emotions like worry or loneliness, and body sensations like sweating or blushing, can bring relief when you've always tried to resist or ignore them. And the value-identification component of ACT gives you an opportunity to contemplate the goals that social anxiety is keeping you from achieving, like being true to yourself, expressing your ideas, or being courageous—and take action to meet those goals.

Furthermore, ACT grants you a different perspective on your social anxiety. It can help you see that your anxiety is not a problem that needs to be eliminated or fixed. It's just an aspect of your human experience. You are not broken because of your social anxiety, you're just stuck. Fortunately, there are innumerable methods to get you unstuck.

As we delve deeper into acceptance and commitment therapy, we'll be exploring these main principles:

Acceptance. In ACT, acceptance is being willing and open to experience the pain and discomfort that we regularly experience as humans, without trying to escape from it. This means making room for all of our emotions, body sensations, or thoughts, including the negative ones that we might normally try to suppress. When we give up resisting or trying to control them, we have more freedom to respond in a positive way.

Cognitive defusion. As humans, we have a tendency to overidentify with our feelings and thoughts. We let them determine our actions, without challenging them or considering their accuracy. Cognitive defusion is the ability to detach from our internal experiences instead of getting hooked by them. It's a practice of simply observing and noticing, allowing our feelings and thoughts to come and go, rather than holding on to or being controlled by them.

Being present. When we dwell on the past or plan for the future, it's easy to get caught up in our thoughts and lose touch with the environment around us. Being present means connecting with our awareness of the here and now, without judging or assessing.

Observing self. This term refers to the process of observing our feelings and thoughts without identifying with them. In ACT, the observing self is often likened to the sky, and our feelings and thoughts are the ever-changing weather. From sunny skies to rainstorms, the weather is constantly changing but does not fundamentally alter the sky.

Clear values. Clarifying our values is the ACT process of reflecting on what matters most to us, deep in our hearts. What do you want your life to be about? What kind of person do you want to be? What do you want to do with your time here on Earth? Clarifying what's important motivates and inspires us, and provides the roadmap for our actions.

Committed action. The "C" of "ACT," committed action is the process of setting goals and taking action to achieve them, guided by our values. Committed action means doing what it takes to live by these values, regardless of what negative thoughts or feelings we're having. Committed action starts with setting goals, identifying obstacles, and persisting flexibly so you can do what matters to you.

How ACT Differs from CBT

ACT and CBT are both powerful, behavior-based therapies that can bring about big life changes. They differ primarily in the view they hold regarding the content versus the context of our unpleasant internal experiences. CBT is effective at helping you identify and change the negative thoughts and feelings that are causing you suffering. ACT, on the other hand, holds that unpleasant feelings and thoughts are natural, and that trying to change them creates more suffering. CBT targets symptom reduction as a primary goal, while ACT acknowledges symptoms may be reduced, but it's considered a secondary by-product. ACT's cognitive defusion and CBT's cognitive restructuring are two different tools that can be applied to change behavior when you're having negative feelings and thoughts. CBT and ACT can work together to help you take effective action and overcome social anxiety.

The First Steps

Social anxiety tries to convince us that we're under constant scrutiny. Isn't it exhausting to be on constant alert for how people may be judging you? The core mindfulness skills of ACT offer ways to work with the challenging thoughts and feelings that come with self-consciousness. Let's break that down.

Defusion

Cognitive fusion refers to the tendency to become caught up in, or buy into, our thoughts. We often listen and believe what our minds tell us, as if it were the truth. When your social anxiety is active, your mind might tell you that you're boring or don't fit in, that speech will be a disaster. What happens when we fuse with these thoughts or beliefs, accepting them as reality? We feel bad and avoid social situations, which pulls us away from living the life we want.

Cognitive defusion, by contrast, is the practice of acknowledging that our thoughts are just that: thoughts. They exist, but that doesn't mean we have to believe them, disprove them, or push them away. Thoughts and feelings come and go like the weather, and like the weather, we can adapt to them. We can't stop the rain, but we can put on a rain jacket or use an umbrella. We can't control our thoughts and feelings, but we can learn to react to them in a more adaptive way until they pass.

Here's a simple cognitive defusion exercise you can try. Suppose you're at a party and have the thought "I'm awkward and nobody wants to talk to me." Replace that thought with this: "I'm *having the thought* that I'm awkward and nobody wants to talk to me." That simple change creates distance between you and the negative thought, helping you stay more objective about it. When you acknowledge a thought in this way, it's easier to understand that your thought is not objective truth.

Another common defusion technique is to imagine a gently flowing stream, with leaves floating on the surface. Visualize placing each thought that comes into your mind on top of a leaf, then watch it float downstream. You're not getting rid of the leaves; you're just watching their natural flow.

When you can recognize your thoughts as thoughts, and unhook yourself from them, you're free to do the things that matter. Dr. Russ Harris, a world-renowned ACT trainer, reminds us in *The Illustrated Happiness Trap* that negative thoughts are "just words and pictures floating through your mind." Being aware of which words and pictures you tend to fuse with in social situations takes time and attention. But with practice, you can get better at catching yourself when you've become attached to your thoughts of self-doubt or self-criticism.

Acceptance

When you're socially anxious, you can often become concerned about anxiety itself. If you're at a social event that triggers your anxiety, at the first sign of anxiousness you might feel scared or on edge, or sense your stomach tighten or get queasy, and try your best to make these feelings go away. You've probably noticed this effort doesn't usually work. One of my favorite metaphors, called "Ball in a Pool," from *The Big Book of ACT Metaphors* by Jill A. Stoddard and Niloofar Afari, explains it this way:

When you try to control your negative feelings, thoughts, or sensations, it's like trying to push a giant beach ball underwater in a swimming pool. You keep trying to hold that ball under the water, out of your consciousness, so you don't have to experience those feelings. The problem is, the ball keeps popping back to the surface. So you have to keep pushing it down, over and over again, or use all of your strength to keep holding it underwater. Struggling in this way is exhausting!

What if you could let go and allow the ball to float to the surface? The ball might float uncomfortably close to your face at first. But with time, you might see it floating on the other side of the pool. The ball doesn't disappear, and you may not like having it in the pool, but if you let it float around without fighting with it, you have your arms free to swim or play. You're no longer preoccupied with getting rid of it. This is what acceptance means in ACT: letting painful feelings, sensations, or urges come and go without a struggle.

With social anxiety, letting go of the ball means giving up your resistance to, and struggle with, intolerable feelings and sensations during social interactions. That might mean letting awkward feelings flow through you while you're making small talk with someone. Or permitting your heart to beat faster as you raise your hand to answer a question in class. By releasing control of the ball, you're developing a willingness to feel all of anxiety's symptoms. As you develop your capacity to open up and accept what's out of your control, you'll find more time and energy to pursue what's important to you. With your hands off the ball, the swimming pool is yours to enjoy.

Being Present

Let's practice being present in this moment, right now. Wherever you are, look around and take in your environment. What do you notice? What colors do you see? Can you feel your body in your seat? Can you feel your feet on the floor? Can you smell anything? What do you hear? If you're willing, put your book down and spend one minute just being in this moment with all of your senses.

Being aware like that, in the moment, without judging or assessing the experience as good or bad, doesn't come naturally to most of us, and takes ongoing practice. Mindfulness exercises that are part of ACT are aimed at helping us to recognize when our mind is wandering and call it back to the present moment. This practice builds our capacity for open receptivity to what's happening to us and in our environment.

When you're in an uncomfortable social situation, it's easy to get stuck in your head. When we're socially anxious, our focus is often pulled toward scrutinizing ourselves, guessing what other people are thinking about us, or worrying about the physical symptoms of anxiety. You find yourself thinking about how you're going to end the conversation or trying to figure out if you're maintaining the correct amount of eye contact. All that self-monitoring leaves little opportunity to pay attention to what's really happening around us. We miss what someone's telling us in a conversation or lose track of what colleagues are discussing in a meeting.

ACT trains you to treat all of those socially anxious thoughts as background noise. We can be aware of them but not focus all of our attention on them. We can bring ourselves away from those distractions, back to the present moment, focusing on the person we're talking with, our surroundings, or the situation in front of us.

Observing Self

Have you ever gazed upon a stunning sea or looked out at a magnificent vista from the top of a mountain and your mind goes quiet? For a moment, you are just silently noticing, perhaps in awe of such beauty. This is your observing self in action. Such moments tend to be brief. It doesn't take long before your thinking self chimes back in with "I need to take a picture," "Where did I put my phone?" and "I hope I brought sunscreen" and so on. As you get hooked by your thoughts again, you are no longer connected with the natural wonder in front of you.

The observing self is the part of our mind that notices or observes our inner and outer world. It's the awareness of your awareness. The counterpart to the observing self is the thinking self. It's the part of our mind that thinks, analyzes, plans, remembers, and so on.

With social anxiety activated, it can take less than a second for your thinking self to disrupt your awareness of a conversation or interaction, piping up and reminding you that everyone is watching you, that you look weird, or that you're probably going to forget what you're going to say. Maybe you're in class and you suddenly realize you've been so caught up in your plans for answering a question if called on that you have no idea what the teacher has been explaining. Or perhaps you're on a dinner date and become so lost in thought about what to say next that you haven't heard a word the other person was saying.

Mindfulness exercises allow us to reconnect with our observing self when it becomes obscured by our anxious thoughts and feelings. If our mind is distracted or lost in thought, with mindfulness we can bring back our observing self. We can simply focus on something and notice that we are noticing.

In Real Life: Sydney's Party

Sydney found throwing parties to be nerve-wracking and stressful. From the moment she sent the first invitation to the morning after the party was over, she was preoccupied with thoughts about being judged. On the day of the event, before the guests arrived, she was nervous with anticipation. Was she dressed appropriately? Was her furniture stylish enough? Was the music okay? Would her food taste good? Would her son behave? The spotlight on her house and her hosting skills felt intolerable, and she hadn't had people over for a special occasion in over 10 years. Sydney sought my help after her son begged her for a party at their house to celebrate his fifth birthday.

Strategy: ACT Mindfulness Skills

Sydney and I worked on ACT mindfulness techniques. By the day of her son's party, she was up for the challenge to try out her new skills. As she was preparing the house for guests to arrive, she practiced accepting her nervousness and the sensation of her heart racing instead of trying to push them away. When her mind drifted to worrying about how things would go, she gently brought herself back to the present moment. Every time she caught a negative thought, such as "My house isn't nice enough," she would tell herself, "I'm having the thought, *My house isn't nice enough*." When jittery feelings surfaced, she let them come and go, without a struggle.

The Outcome

Sydney's stress didn't completely disappear, but she felt calmer and more confident before and during the party than she had in years. By defusing from her thoughts, she was able to see, for the first time, how unrealistic her expectations for hosting had been. When she gave up trying to control her anxious feelings and bodily sensations, and let them be, she experienced more ease. By learning to be present, she was able to deeply enjoy many moments during the party. Her most cherished moment was watching her son brim with joy while everyone sang "Happy Birthday" to him in their backyard.

DEFUSION

The aim of defusion is to loosen your attachment to a negative thought, disentangle you from it so you don't bite the hook and believe the thought to be the truth. Here's an easy exercise you can use to avoid getting entangled with thoughts that cause social anxiety. It's adapted from a technique created by ACT founder Steven C. Hayes.

How to do it:

1. Think about the messages you tell yourself when you are feeling self-conscious or anxious in a social or performance situation. Write down five self-critical thoughts.
   Examples: *I'm a loser. My [body part] looks weird. I'll sound dumb. I'll make a mistake.*

2. Choose one thought from the list to work on. Rewrite that thought by preceding it with "I'm having the thought that..."
   Example: *I'm having the thought that I'm a loser.*

3. To create more distance from the thought, rewrite the previous thought, adding "I notice I'm having the thought that..."
   Example: *I notice I'm having the thought that I'm a loser.*

4. Repeat steps 2 and 3 for the remaining four self-critical thoughts.

Did your perception shift? Were you able to defuse from your thoughts? Write down your reflections.

You can practice defusion anytime you catch a negative thought while in the midst of a situation that's triggering your social anxiety. It can also help you after a social interaction that didn't go as well as you had hoped.

NOTICE FIVE THINGS

Here's a simple, 5-minute mindfulness exercise to help you slow down and be present by tuning into your surroundings. You can practice it anywhere throughout the day, especially when you find yourself feeling anxious or self-conscious. You can also use this exercise to transform any routine activity into an opportunity to practice being present in your environment.

How to do it:

1. **Breathe.**

   Sit upright with your feet planted flat on the ground.

   Rest your hands on your thighs.

   Take three slow, deep breaths.

   Slowly look around your environment.

2. **Engage.**

   Notice five things you can SEE.
   Examples: *people, cars, buildings, trees, sky, clouds, colors, shapes, textures*

   Notice five things you can HEAR.
   Examples: *cars, talking, birds, leaves, airplanes, appliance humming*

   Notice five things you can FEEL.
   Examples: *Your back in your chair, your watch on your wrist, your pants on your legs, air on your face, feet on the floor*

3. **Reflect.**

   Take three more slow, deep breaths.

   How do you feel now? Compare how you feel now with how you felt 5 minutes ago. What has changed?

Values-Driven, Committed Action

The ultimate goal of ACT is for you to take action based on the values that are most important to you in order to live a rich, intentional life. When we're socially anxious, we're hooked by our negative thoughts, and we let them convince us that we should avoid people. But avoidance takes us further away from doing what's important to us, like developing relationships, having a social network, or expressing our true selves. The disconnect between our intentions and our actions can lead to us feeling isolated, lonely, depressed, or even guilty, which makes it even harder to reach out. We may want to change, but as Pablo Picasso once said, "What one does is what counts. Not what one had the intention of doing."

"Passengers on the Bus" is one of ACT's most popular metaphors, developed by Steven C. Hayes. Imagine you're a bus driver, and the bus you're driving is full of passengers. While you're driving, the passengers get angry and upset. They point out what a lousy driver you are. They yell out their demands: "Turn here! Don't go there! Pull over!"

You try desperately to get the passengers off the bus, but they won't budge. So to satisfy them, you adhere to their demands and let them tell you where to go. Which is usually the same, comfortable road, over and over again. Eventually, however, you realize that you're the driver, and you get to decide which direction the bus goes. Since the passengers won't get off the bus, they'll just have to come along for the ride.

The passengers are your thoughts, feelings, and bodily sensations. They're on the bus with you, and you can't get them off, and they frequently bombard you with demands and complaints. But you're the one in the driver's seat, and you don't have to go where they tell you to.

And that's the concept of values-driven, committed action. It means doing what it takes to move persistently in the direction of your values, even in the presence of obstacles, discomfort, or pain that demands you change course. We're going to spend the remainder of the chapter working on ACT skills that clarify the values that matter to you, help you create goals guided by these values, and support you in taking action to achieve those goals.

Establishing Clear Values

> "Values are what you want your life to be about, deep in your heart. What you want to stand for. What you want to do with your time on this planet . . . What you would like to be remembered for by the people you love." —Dr. Russ Harris, *The Happiness Trap*

Getting clear about what matters is essential for guiding your actions toward what your heart wants. And it all starts by knowing your *values*—that is, the qualities of behavior that reflect who you want to be and what you want to stand for. If your values are clear, it becomes much easier to choose where to invest your time and energy. If you're unsure, here are some questions to consider:

- What really matters to you?
- How do you want to behave?
- What kind of person do you want to be?
- What do you want to stand for in the world?
- What do you want to spend your time and energy doing?

In ACT, values are often referred to as "chosen life directions." They're like a compass, because they give us direction and guide us on our ongoing journey. Your values aren't an objective to be achieved; rather, they're a direction to follow. We might use a compass to travel east, but no matter how fast or long we travel, we don't arrive at a final destination called "east." We can always go farther, and following our values works the same way.

I came up with my "east" in a workshop a while back. We were invited to create our life charters. After a lot of thought, I realized that what lights me up is being a stand for people in the world brimming with love and joy. That continues to guide the way I live my life.

Setting Goals

Suppose you want to be in a loving, intimate relationship. That's a value. It involves ongoing action: giving and receiving love and affection or being vulnerable. If, as you follow this value, you want to ask someone on a date or get married, those are goals. You can check them off a list. If you want to be your true, authentic self at work, that's a value. At no point would you be "done" being your true self. If you want to speak up honestly in a meeting, that's a goal that can be achieved.

A goal can be achieved, whereas values are ongoing. If our values were a compass guiding us north, a goal would be a hill we wish to climb while heading northward. Once your values point you in a certain direction, it becomes clearer what goals you'll need to accomplish during your journey.

The ACT model categorizes goals as immediate, short term, and long term. An immediate goal is something you can easily accomplish within 24 hours. A short-term goal is action you can take within a few days or weeks. A long-term goal might take months or years to achieve and could require multiple actions and a detailed plan.

ACT encourages setting goals that are SMART:

Specific: What specific actions will you take?
Meaningful: Does it matter to you?
Adaptive: Will it improve your life?
Realistic: Is this goal achievable?
Time-framed: Did you set a day and time when you will do it?

Example: After considering how your social anxiety impacts your life, you might choose to focus on relationships. You identify "Being courageous in my interactions with others" as an important value to you.

An immediate goal could be to ask, "How's your day going?" to the barista at the coffee shop the next morning.

A short-term goal might be to smile and say "Hi" to one new person each day for a week or two.

A long-term goal might be to join a club or sports league, or cultivate two friendships by the end of the year.

In each case, the goal is specific, meaningful (you find these things challenging to do, so it will be a victory to accomplish them), adaptive (your life will be improved if you make new friends), realistic (you're starting with easier goals to build confidence), and time-framed.

Example 2: Let's say you value engaging in fun-filled activities with others.

An immediate goal might be to ask a friend to go on a hike today.

A short-term goal could be to try out a rock-climbing class within the next two weeks.

A long-term goal might be to go on a vacation with a friend within a year.

Taking Action

If you know where you want to go but don't take steps to get there, the knowledge won't make much of a difference in your life. At its core, the focus of ACT is on taking action—walking the walk, as Steven C. Hayes refers to fulfilling goals.

Taking action means looking at your goals, both short and long term, and focusing on realistic, concrete steps you can take to achieve them. Let's say one of your goals is to make a new friend. A first step might be to choose one person you already know, whom you'd like to become closer to. Next, you might call and ask that person to coffee. After that, your next step could be to invite the person to lunch. You could work your way to spending a day together doing something you both enjoy, like going to the beach. Repetition like this is what builds a friendship.

Breaking big goals into smaller, more manageable steps is not only practical, it can mitigate feelings of being overwhelmed. If you say to yourself that you want to find a romantic partner, you might be paralyzed by the magnitude of the task. Instead, you can ask yourself, what steps are involved in finding someone? You can make the steps as granular as you need to so each task feels manageable:

1. Register for an online dating app.
2. Choose one picture of myself.
3. Spend 10 minutes looking at potential matches.

After a task has been broken down into a series of smaller steps, the next move is to decide on a specific time frame for completing each one. I have my clients put their tasks on the calendar on their phones, set with an alert. Scheduling your action steps makes them feel real and more easily achievable.

Let's see how the ACT approach to goal setting and taking action can work.

In Real Life: Aaron's Reading

Aaron sought my help after he was asked to do a reading during his twin sister's wedding ceremony. He had panicked and said "no," then immediately felt guilty. Aaron had always felt uncomfortable when he felt there was a spotlight on him. He tended to overestimate how much other people noticed him, and often felt overcome with self-consciousness and embarrassment. He was convinced people were making fun of or pitying him behind his back.

The Strategy: ACT: Values-Driven, Committed Action

While we explored his values, Aaron figured out what was most important to him: spending time with family and being courageous. He set some specific goals to translate these values into committed action. His first was to call his sister to let her know he had changed his mind and would do the reading after all. Participating in the wedding would guide him in the direction of both of his chosen values. For the following months, he practiced ACT mindfulness skills to make sure his negative thoughts and feelings wouldn't sway him from spending time with family and being courageous.

The Outcome

On the day of his sister's wedding, Aaron felt panic creeping back, but he reminded himself why he was taking such a big risk: to celebrate with his family and grow as a person by being courageous. As the ceremony began, Aaron was able to stay present by pulling his attention toward the guests, the beautiful decorations, and the sounds of people chatting happily around him. During his reading, Aaron felt uncomfortable, but when he looked out at the guests, he noticed a sea of friendly faces smiling back at him. It was both shocking and exhilarating to realize they weren't secretly waiting for him to fail but sincerely wanting him to succeed. When he glanced at his sister, Aaron saw her smiling with joy and felt proud of his courage.

VALUES ASSESSMENT AND GOAL SETTING

Your values are what are most important to you in life. They help to determine life priorities and influence your actions. This exercise walks you through assessing your values and creating goals that you can act on to support those values.

How to do it:

PART 1: VALUES ASSESSMENT

1. Choose one area in your life that is important to you where social anxiety is having a significant negative impact:

Intimate relationships	Family relationships	Recreation/fun
Friendships/social life	Career/employment	Learning/growth

2. Answer the following questions to help you clarify what values you'd like to bring to this one area of your life:

   What matters most to you?

   Which of your positive qualities would you like to build on?

   What qualities would you like to bring to your relationships? What kind of partner/friend/worker do you want to be?

   How do you want to interact with others?

   What kind of difference do you want to make in this area with your limited time on Earth?

Here is a list of possible values or "ways of being" to help you generate your own list.

Adventurous	Easygoing	Open
Authentic	Forgiving	Patient
Brave	Fun	Peaceful
Calm	Genuine	Playful
Connected	Graceful	Powerful
Courageous	In action	Self-expressed
Creative	Kind	Understanding
Curious	Loving	Unstoppable

3. Consider your answers, then write down one value or way of being you would like to focus on.

PART 2: GOAL SETTING

Your goals will give you a practical way to put the values you listed earlier into action. They are practical, achievable events that move you in the direction of a life you love.

In your notebook, write down three goals you would like to achieve, related to the value you chose to focus on. Remember to make your goal a SMART one.

Immediate Goal (within 24 hours)

Short-Term Goal (with the next few weeks or months)

Long-Term Goal (within a year)

COMMITTED ACTION

To make a real difference in your life, action is needed. Committed action is about maintaining motivation to continue in the direction of your values over time. Let's determine some actions that will bring your goals to fruition.

1. Copy the short-term goal you wrote down in the previous exercise.

   Example: *Build one new friendship.*

2. Write down three specific steps you could take to achieve this goal within the next weeks. Be specific.

   Examples: *Call Miriam. Go to the upcoming networking event. Email Paul asking him to be a reference.*

3. A common reason why we fail to achieve our goals is that we tend to get discouraged by obstacles and challenges. These barriers can be our old ways of dealing with social anxiety, like avoiding uncomfortable social situations or letting our negative thoughts stop us. Choose one of the specific actions you wrote in step 2. List three potential barriers you might expect in trying to enact that action.

   Examples: *She won't want to hear from me. I'll cancel at the last minute. I'll be too nervous to meet.*

4. Now go back and write down one solution for each potential obstacle that you recorded. Think about each potential snag and come up with something you could do or say or tell yourself to get past it. What could you do? What could you tell yourself?

   Examples: *I'll remind myself that we've had good conversations in the past. I'll go there early so I'll have time to get calm and comfortable. I'll use mindfulness to detach from my nervous thoughts.*

5. Research shows that if you share an intended action with at least one other person, you are more likely to feel accountable and follow through. Choose one person in your life with whom you'll share some of your goals and steps.

CHAPTER FIVE

Cultivate Calm

"You can't stop the waves, but you can learn to surf." That's a well-known saying by world-renowned mindfulness teacher Jon Kabat-Zinn, describing his teaching of mindfulness.

In life, we can't possibly control the waves that come our way, and yet we spend so much of our energy trying. But like the surfer, we do have control over how we interact with them. We can fight or brace against the surges and swells. Or we can use mindfulness, a quality of being that we can bring to anything we do, to surf that wave wherever it takes us and not get pulled into the undertow.

When we're in a social situation that triggers our fears, with mindfulness we can ride the difficult feelings and thoughts by staying fully present with what's happening inside and around us. We accept our body's stress response, and our anxious feelings or thoughts, without judgment or resistance. Being present to what's going on in the moment, without labeling it as bad or good, allows us to slow down and bring inner calm to the situation.

In this chapter, we'll explore ways we can grow our capacity to dwell in the present moment through mindfulness and meditation. We'll learn exercises that make it easier to respond to stressful situations with less reactivity and anxiety, and more awareness and calm. And when you're riding that wave, instead of being pummeled by it, you'll find that social interactions are actually pretty fun. Just like surfing.

Mindfulness and Meditation: A Primer

We live in a time when Eastern philosophy has become part of mainstream Western culture. You've probably heard the words "mindfulness" and "meditation" many times; they've become everyday terms. But if you're not involved in these practices, you may not be certain how they relate to each other. Mindfulness and meditation share many similarities, and their practical applications can overlap, but they are not the same.

The current wave of mindfulness as therapy has its roots in a stress-reduction program developed by Jon Kabat-Zinn, PhD, a professor of medicine at the University of Massachusetts. In 1979, he adapted Buddhist teachings on mindfulness and developed an 8-week course called Mindfulness-Based Stress Reduction (MBSR). Hundreds of hospitals, clinics, and mental health centers around the world now offer MBSR to help people relieve stress, anxiety, depression, chronic pain, and other conditions.

As we touched on in the last chapter, mindfulness is the ability to be fully present, in the here and now, aware of where you are and what you're doing. Whenever you bring your full awareness to what you're experiencing, you're being mindful. Which means you can practice mindfulness anytime and anywhere: while walking, while meditating, while washing the dishes. Simply by paying attention and being present in whatever you're doing, with all of your senses, you're engaging in mindfulness.

Here are four important points to know about bringing mindfulness into your life:

It's more challenging than you might think. It's easy to say, "pay attention to the present moment," but not always easy to do it. Our minds naturally wander; we're always getting lost in thought. One study found that people spend almost half of their days thinking about something other than what they are doing. Most of the time, we're dwelling on what's happened in the past or worrying about what might happen in the future. Which is all to say: It's difficult for the human mind to stay in the present moment, so don't be upset with yourself when you struggle to be mindful.

Mindfulness is a skill. We all have the natural ability to be mindful at any moment, but this ability becomes more readily available when we practice.

What's nice is that you can practice mindfulness informally, in any situation. You can be mindful while you work, during a conversation with someone, or in a meeting. It's a quality of mind you can bring to anything you're doing.

You'll get better with a formal practice. Along with bringing mindfulness to the informal moments of your day, you can strengthen and support your ability to be mindful with a committed, formal practice. This is where meditation comes in—it's the most common and accessible formal practice of mindfulness. We'll be sharing some mindfulness meditation exercises in this chapter. Mindful meditation is typically done while sitting, but some variations combine it with movement or exercise, like tai chi or walking.

Mindful meditation is exploration, not a destination. When we meditate, we bring a curiosity to the workings of our minds using a variety of techniques: focusing on the breath, counting inhales and exhales, repeating a mantra, conducting body scans, visualizing our thoughts, or just letting thoughts pass without judgment or attachment, like birds in the sky.

Benefits of Mindfulness

Being able to calm your anxious mind with mindfulness can be an empowering tool for overcoming social anxiety. When you pay attention to what's happening in your body during anxiety, you manage your stress response and increase your tolerance for discomfort. You can slow down your response to social triggers, giving you a pause to stop your mind from spiraling into worry, and respond in a different way.

Its roots may be in the venerable traditions of Buddhism, but modern science backs up the benefits of mindfulness and mindfulness meditation. For example, in a study done at Stanford University, participants who had social anxiety disorder underwent an 8-week MBSR program. Afterward, they reported less anxiety and more positive self-esteem. Here are some documented benefits that are particularly relevant for counteracting social anxiety:

Improved emotional regulation. Mindfulness and meditation provide the tools needed to step back from intense negative emotions and accept them instead of trying to escape or struggle with them. This practice allows you to better regulate your emotions, reducing anxiety in social interactions and increasing overall emotional stability.

Increased emotional intelligence. The ability to recognize your own emotions, and those of others, and manage them effectively in your relationships is called emotional intelligence. During meditation, you're able to get in touch with your emotions, recognize them as fleeting, and learn to be present with them instead of being afraid to face them. This improves how you react, internally and externally, during social situations that cause anxiety.

Heightened focus. Meditation and mindfulness can strengthen your attention span and increase mental clarity and focus. When you're focused on the present moment, you are fully engaged in what you're doing, without getting caught up in the chatter of your mind or mind-wandering. That means you're not believing the story your inner critic is spinning about how a social interaction could go wrong.

Reduced stress and more relaxation. Stress reduction is one of the most common reasons people try meditation. Who doesn't want less stress? Meditation affects the body in the opposite way that stress does—by activating the body's relaxation response, the antidote to the fight-or-flight reaction. It calms your mind and body by quieting, and distancing you from, thoughts that keep your body's stress response activated. Many people who meditate regularly have learned to relax their bodies on demand, which allows them to more effectively manage stress in the moment.

Increased empathy. One particular form of mindfulness meditation, known as loving-kindness meditation, is a compassion-based meditation practice that focuses on developing feelings of kindness and warmth toward others. Regularly practicing this type of meditation exercise strengthens areas of the brain responsible for empathy and kindness. It also appears to enhance helping behaviors toward others. People who practice loving-kindness meditations report more positive emotions and increased social connection. This kind of meditation can also reduce self-criticism and improve self-compassion.

Improved relationships. Meditation has been shown to strengthen relationships and improve your ability to relate to others. It sharpens your ability to pick up on cues about how others might be feeling. Mindfulness meditation also helps you be a more attuned listener and communicator, which enhances any relationship and makes social interactions smoother.

Increased happiness. Meditation can increase your happiness by releasing "feel-good" chemicals into your brain and nervous system, like serotonin and endorphins. Studies have shown that meditation also increases the release of dopamine, a chemical associated with feelings of pleasure. A dose of feeling good can assist in soothing awkward or uncomfortable feelings that might come up in social situations.

LOVING-KINDNESS MEDITATION

Loving-kindness, or goodwill, comes from a sincere wish for ourselves and others to experience well-being. The loving-kindness meditation, also called *metta*, is the practice of directing kind wishes toward others. We start with ourselves and gradually extend our well-wishes for well-being to all beings on the planet. The following loving-kindness meditation is a modified version based on classic metta practices.

How to do it:

1. Sit in a comfortable position.
2. Take three slow, deep breaths. As you breathe, imagine your breath moving though the area of your heart.
3. Sitting quietly, mentally repeat the following messages for one minute:

   May I be happy. May I be peaceful. May I be at ease.
4. Allow yourself to bask in any feelings of warmth and compassion toward yourself for a few moments. If your mind wanders, gently bring it back to feelings of loving-kindness toward yourself.
5. Now bring to mind the image of a person close to you whom you love and care about deeply. Visualize this person in front of you as you repeat the following loving-kindness messages for one minute. As you repeat these words, connect with any feelings of gratitude or love that might be present when you're thinking about this person.

   May you be happy. May you be peaceful. May you be at ease.
6. You can bring to mind other important people from your life, one by one, and repeat the messages for each of them.
7. You can extend feelings of goodwill to include groups of people and even everyone around the world. You could send kindness to a stranger, or even a person you don't like. Repeat the following messages for one minute.

   May all beings be happy. May all beings be peaceful. May all beings be at ease.

Mindfulness and Self-Compassion

Most of us would agree that compassion is a virtue. But often we forget to include *self-compassion* in that characterization. Bringing warmth, tenderness, or friendliness to ourselves when experiencing the pain of being socially anxious can be difficult, since people with social anxiety tend to be self-critical. But replacing that criticism with self-compassion is important, because whatever negative thoughts you have about yourself, you'll likely imagine others are thinking the same things about you. Learning to be self-compassionate can shift the way we view and treat ourselves when we're suffering from social anxiety.

Kristin Neff, PhD, a pioneer in the field of self-compassion and author of *Self-Compassion: The Proven Power of Being Kind to Yourself*, describes three key elements to self-compassion:

Self-kindness. Instead of beating ourselves up mentally, we're warm and understanding toward ourselves. Self-compassionate people recognize that being human means being imperfect, and making mistakes is expected.

Common humanity. Instead of thinking you are the only person suffering or making mistakes, you recognize that all humans suffer, and that feeling inadequate is part of our shared human experience.

Mindfulness. When we are learning to be in the present and accept our painful thoughts, feelings, and sensations instead of denying or fighting them, there's room for us to feel compassion for our pain.

Just like other mindfulness practices, self-compassion is a skill that can be strengthened over time. Pema Chödrön, a Buddhist teacher, international lecturer, and best-selling author writes about the Buddhist concept of *maitri*, which means unconditional friendliness or *loving-kindness* that we can cultivate toward ourselves and others. Making friends with yourself means embracing the parts of yourself that you don't like or that you wish you could get rid of. Just like a true friend, you wouldn't turn your back on or abandon yourself when your darker side shows up.

When you can become friends with your mind and body, it becomes more comfortable to stay in the present moment, in all situations, without beating yourself up.

SELF-COMPASSION EXERCISE

Practicing self-compassion means being kind and understanding toward yourself when you fail, when you think you're going to fail, or when you're focused on something negative about yourself. When this happens, think about how you might treat a close friend or loved one having a hard time, and acknowledge that making a mistake, experiencing difficulties, or falling short of your expectations are all part of what makes you a human being. This self-compassion exercise is adapted from the Mindful Self-Compassion program, developed by Christopher Germer, PhD, and Kristin Neff, PhD. You can use it anytime, especially when you become aware that you are criticizing yourself.

How to do it:

1. Write down a social situation that's causing anxiety in your life right now.
2. Imagine the situation vividly in your mind. Experience it as if it were happening right now. Where are you? Who are you with? What is happening?
3. Note the anxiety or discomfort that you're feeling. As you put yourself into the scene, where do you feel anxiety in your body? Experience it without passing judgment.
4. Say to yourself: "This is a moment of suffering. This is anxiety."
5. Connect with humanity as you say to yourself, "Other people with social anxiety feel this way. I'm not alone."
6. Put your hands over your heart and say to yourself, "May I be kind to myself."

Mindful Nonjudgment in Action

While the time we spend in meditation does make a positive impact on the quality of our day and increases our capacity for mindfulness, being mindful is not limited to a formal practice. It is available during ordinary day-to-day activities, such as doing the dishes, folding laundry, or brushing our teeth. In fact, anything we do can instantly become an opportunity to practice mindfulness. You can turn ordinary tasks into mindfulness sessions by paying attention to any physical sensations you're experiencing. You can also practice letting your feelings and thoughts be present, without attachment or judgment, while you're performing a routine task.

If you're doing the dishes mindfully, for instance, instead of rushing through it to be finished or doing it while thinking about something else, you can focus on your senses. What do you notice about the water running over your hands or the smell of the soap? Wash each plate with intention, concentrating on your breathing, and let go of thoughts that come up. When you observe that you're lost in thought again, just notice it, and come back to your breath or senses.

Applying the principles of meditation and mindfulness to your daily activities can keep you tuned in to the moment, yourself, and the environment around you. Pausing to practice mindfulness for just a few minutes throughout the day, each day, can reduce your anxiety and grant you a greater sense of calm and ease.

Here are some more options for bringing mindfulness into your day.

Mindful Breath Meditation

Our emotions, particularly intense emotions like self-consciousness and fear, can obscure the full picture of what's actually happening within ourselves and around us. Mindful breathing is a fundamental technique for countering this by simply focusing your attention on your breath. Pay attention to the inhalations and exhalations, feeling the sensations, without trying to adjust or control them. That's it! This mindful breath meditation is helpful when you're feeling fearful or anxious. Practicing regularly can make it easier to use in the moment when you need it in a difficult social situation.

It's easiest to do a mindful breath meditation while sitting, but you can also practice it standing or lying down. Your eyes can be open or closed. The technique is simple: Breathe normally, and simply observe each breath without trying to adjust it. Focus on the contraction and expansion of your abdomen. You may find your mind wanders; this is natural, and not a problem. Just notice when this is happening, say "thinking" to yourself, and without judging, gently bring your attention back to your inhaling and exhaling.

Mindful Listening

When we listen to another person, our mind naturally drifts. We can get distracted by our own thoughts and worries and fail to hear or remember what the other person is saying. When we're socially anxious during a conversation, we might be so engaged in trying to figure out what they're thinking about us, or planning what we're going to say next, that we don't catch what's being said to us, making an uncomfortable experience worse.

The next time you're with a friend or loved one, you can practice being mindful while they speak. Focus all of your attention on the other person, and let your thoughts and worries come and go without attaching to them. You can also focus on your breathing during the conversation, to anchor you in the here and now. When you catch your mind wandering, gently bring it back to the present and focus fully again on the person in front of you. As this becomes familiar to you, you'll be able to call on mindful listening during social interactions that trigger your anxiety.

In addition to reducing social anxiety and increasing feelings of calm, a habit of mindful listening promotes effective communication, empathy, and understanding. Listening can be an act of love and kindness that deepens your connection with another person.

Mindful Walking Meditation

Whether you're walking down the street, in the woods, or down the hallway to a meeting, mindful walking is an opportunity to guide your mind out of autopilot mode and bring your attention to the present moment. You can turn walking into a mindful exercise almost anywhere or anytime.

Start by walking slowly and deliberately. Notice how the ground feels under your feet. Notice the movement in your legs or the way your arms sway. Look around and pay attention to the details in your surroundings, like the plants and the trees, the color of the sky, or the architectural details of the buildings you're passing. You can count steps, too, if that's helpful.

Be present in the here and now with every step you take. When your mind wanders, guide it back again by returning your attention to the sensations of walking, noticing your feet touching the ground again. When you're ready to end your mindful walking meditation, stand still for a moment and consider how you might bring this awareness to the rest of your day.

Mindful Eating

Social events often include eating, which can be stressful for some people. You might be suffering through meals with others due to your social anxiety. It's easy to eat mindlessly when you're focused on others, trying to avoid looking weird, or worrying that you might spill something and embarrass yourself. Calling on mindfulness when you eat, as in other situations, separates you from the negative thoughts and feelings of social anxiety.

A meal or a snack includes so many sensory experiences, it's a prime opportunity to practice focusing your full attention on the present moment. Eating slowly allows you to engage your senses and become aware of the textures, flavors, aroma, and other qualities of food, appreciating your meal in the process.

Begin by sitting comfortably. Focus your attention on your immediate experience. Before you eat, look at your food and take notice of the colors and textures. Observe any aromas. Take small bites and eat slowly. How many different flavors can you detect in each bite? How does the food feel in your mouth? What are the sensations as you chew and swallow? What about the sound of the spoon against the bowl, or the texture of sandwich bread on your fingertips? At what point do you start feeling full? Eating slowly and attending to your senses allows you to be fully present to savor the flavors, aromas, and textures.

During your next meal with others around, you can practice mindful eating by shifting your attention to the experience of eating. Notice how good the food tastes in your mouth. Let your slow chewing and swallowing bring you back to the here and now. Then shift back to the conversation, listening mindfully to the people at your table, then back to mindful eating. Repeat this process throughout the meal to ease self-consciousness and bring more calm to your experience.

Mindful Body Scan

The body scan is a foundational mindfulness practice that attunes you to physical sensations, bringing mindful attention to what the body is experiencing. Research has shown that stress reduction is one of the primary benefits of mindful body scans.

As the name implies, body scanning involves paying attention to parts of your body in a gradual sequence from feet to head. Along the way, you'll notice and release tension, making this practice an effective tool for helping you relax when social anxiety is triggered. Body scans also train you to tolerate unpleasant sensations by teaching you to simply notice them without trying to change anything. This is especially helpful with the physical manifestations of social anxiety—racing heart, sweating, feeling dizzy—which can be difficult sensations to accept in the moment.

You can practice a body scan anytime you feel stressed or anxious, or do it once a day as a regular practice. Start by lying down and allowing your breathing to slow naturally as you pay attention to your inhales and exhales. First bring awareness to your feet, observing any sensations or discomfort, and simply notice any thoughts or feelings that come up. As you breathe, imagine your breath traveling into any areas of discomfort or tightness. Visualize tension leaving your body as you exhale. Continue this practice with each area of your body, gradually moving up from your feet to the top of your head.

In Real Life: Adela's Job Interview

Adela was a client who came to see me for help with job interview anxiety. When she first moved to San Francisco 10 years ago, she'd been thrilled to get a computer engineering job at a promising startup. Over the last year, however, she'd become bored with her position and was interested in joining a new company. She wasn't applying for jobs, though, because she'd been suffering from intense anxiety related to job interviewing. Her last job interview was a nightmare. As much as she tried to stay calm, the anticipation and experience of being assessed in the spotlight led to a hasty escape to the bathroom due to severe nausea. The worst part was the feeling of unreality, or derealization, she experienced during the interview. Adela had the sensation that things around her weren't real, which made her feel even more anxious.

Strategy: Mindful Breath Meditation

Adela noticed an open director of engineering position and was eager to get a handle on her job interview anxiety. We tried out a variety of mindfulness techniques, and she found the mindful breath meditation to be the most helpful. Adela practiced it regularly and even downloaded a mindfulness app on her phone so she could practice anywhere.

At first, Adela could only focus her attention on her breathing for one minute at a time. Her mind would naturally wander, but she practiced gently bringing her focus back to her breathing. After a few weeks, she was able to practice the mindful breath meditation for 20 minutes.

The Outcome

Adela decided to send her résumé for the director position she was interested in. When she was called for an interview, she felt both excited and nauseous. Adela arrived for the interview early and started her mindful breath meditation as soon as she sat down in the waiting room. She could feel some queasiness in her stomach, and her mind was racing, but she continued to pull her attention back to her breathing, as she had been practicing. Her jitters subsided to a tolerable level, and Adela felt a subtle shift as her sense of calm took hold. Throughout her interview, she was able to gently bring her focus back to her breath. At the close of the meeting, Adela was composed and relieved when they asked her to come back for a second round.

PROGRESSIVE MUSCLE RELAXATION MEDITATION

Progressive muscle relaxation is an exercise that reduces anxiety in your body. You will be slowly tensing (but not straining) and then relaxing muscle groups throughout your body. This exercise can be used to provide an immediate feeling of relaxation in the body when you need it, or as a daily practice to regularly release tension you may not even realize you're experiencing.

How to do it:

Each muscle should be tensed for 5 seconds but not strained or tensed to the point of pain, then relaxed for 10 seconds before you move to the next muscle group. If you have any pain or injuries, you can skip the affected areas.

1. Sit or lie down, with eyes open or closed. Begin by taking a few deep breaths. Then focus your attention on your hands and arms. Tense the muscles of both your hands and arms, holding them tense for 5 seconds. Release the tension in your hands and arms and relax for 10 seconds.
2. Focus your attention on your eyes and cheeks. Tense the muscles of your eyes and cheeks for 5 seconds, as if squinting. Release and relax for 10 seconds.
3. Next, focus your attention on your mouth and jaw. Tense your lips and jaw for 5 seconds. Release and relax for 10 seconds.
4. Next, focus your attention on your shoulders. Tense your shoulders for 5 seconds. Release and relax for 10 seconds.
5. Next, focus your attention on your chest and stomach. Tense your chest and stomach for 5 seconds. Release and relax for 10 seconds.
6. Next, focus your attention on your legs. Tense your upper and lower legs for 5 seconds. Release and relax for 10 seconds.
7. Next, focus your attention on your feet. Tense your feet for 5 seconds. Release and relax for 10 seconds.

Mindfully Grateful

When you practice mindfulness, it's common to experience more gratitude: The more we slow down and pay attention to the present moment, the more we are able to appreciate the good things in life. Mindfulness allows us to become present to what's in front of us, from our surroundings to the people in our lives, and out of that grows a natural spirit of thankfulness. You can't feel grateful for things you don't notice, and so mindfulness and gratitude are intertwined.

Gratitude is greatly beneficial for social anxiety because of the way it affects the parts of the brain that regulate emotions and help with stress relief. The Mindfulness Awareness Research Center of UCLA found that gratitude actually changes the neural structures of our brains, making us happier and more content. Gratitude can reduce the levels of stress hormones in our body, which in turn reduces symptoms of anxiety while increasing our capacity to handle stress more effectively. And it stimulates our body to release the chemical messengers serotonin, norepinephrine, and dopamine, which boost our mood and help us manage anxiety.

Here are some other ways that gratitude frees you from social anxiety:

Gratitude changes how you see yourself and the world. Turned toward one's self, gratitude can include appreciating your own good qualities and recognizing what's great about you. This can increase your confidence and allow you to feel more comfortable around others. Many studies have shown that people who have a gratitude practice experience higher levels of self-esteem. Turned outward, experiencing and expressing gratitude can encourage you to focus on the positive things in your life, which naturally improves your mood and outlook on life in general.

Gratitude builds and sustains friendships. Several studies have found that thanking a new acquaintance or acknowledging a person's contributions makes the person more likely to seek a more lasting relationship with you.

Gratitude facilitates optimism. As you rewire and train your brain to notice the good and be grateful, you naturally become more optimistic. Social anxiety often manifests as overestimating the likelihood that something negative will happen and underestimating your capacity to handle it. Focusing on the positive in ourselves, others, and the world around us can help alleviate this kind of thinking. An optimistic person can recognize negative things in life but chooses not to dwell on them. For example, an awkward social encounter or a problematic performance can feel like a temporary setback rather than something to despair over.

Starting a gratitude practice is easy, and you can experience immediate benefits. Paying attention to what you are grateful for becomes easier as you practice it; studies have shown that it takes only 10 weeks of practicing gratitude to experience a drastic increase in optimism. While shifting your mindset in this way won't turn you into a Pollyanna (thankfully), it does offer the benefit of increasing your well-being and life satisfaction.

A gratitude practice can be as simple as taking the time to notice and reflect on things, people, or situations for which you're thankful. You can make this a gratitude meditation by visualizing all the things in your life for which you're grateful, and perhaps adding a mindful breath exercise. You can practice gratitude meditation anywhere—sitting on a cushion in silence or while waiting for your morning coffee to brew. At its core, this practice just means reflecting on what you are thankful for in your life.

Here are six other ways you can practice gratitude:

1. **Keep a gratitude journal.** One of the most effective ways to benefit from gratitude is to turn it into a habit. Writing down a few things you are thankful for each day is one of the easiest and most popular exercises to nurture gratitude. Psychologist Robert Emmons, PhD, author of *Thanks! How the New Science of Gratitude Can Make You Happier,* explains that keeping a gratitude journal, in which you write brief reflections on moments you're thankful for, can increase your happiness by as much as 25 percent.

   Designating a specific daily or weekly time for this will help your gratitude journaling become part of your normal routine. The method is to reflect on the past day or week, remember a few things for which you're especially grateful, and write about them however you see fit. For a more formal exercise, there are countless gratitude prompts available online that offer suggestions of what to write about in your gratitude journal.

2. **Write gratitude letters.** Think of someone in your life who gave you unconditional support or made a big, positive impact. Write this person an email, letter, or card, describing what they did that makes you feel grateful. Share how their help influenced your life for the better. Be as detailed and specific as possible. You can also express your appreciation for this person's positive qualities, or any memories you treasure. If you have time and are able, personally delivering this kind of letter can be a transformative exercise.

3. **Keep a gratitude jar.** This simple practice can have a deep impact on your daily well-being. Each day, you can write one to three things that happened that day that you're grateful for. It can be something small, like getting to your train on time, or something as big as the love you feel for a friend or family member. Write a summary on a slip of paper and place it in a glass jar or other transparent container that you keep where you'll see it every day. As the jar fills, you'll be replete with many reminders about what is good in your life, which you can review whenever you want a gratitude boost.

4. **Keep a gratitude rock.** Pick a rock you like. Maybe it's smooth, it has a compelling texture, or you found it in a special place. This is now your gratitude rock. Carry it in your pocket, leave it on your desk, or put it in a special place where you'll see it every day. Each time you see it or feel it in your hand, pause and bring attention to one thing you are grateful for in that moment. The rock can also spark a mindfulness moment in your day, bringing you into the present, and reminding you to be aware of your surroundings.

5. **Take a gratitude walk.** Take 20 minutes to walk outside by yourself or with someone else. As you walk, notice 10 things around you for which you can be grateful. For example, you could notice a tree and generate gratitude for what trees provide for you in your life, from the shade they provide to the paper you write on. You can express your gratitude silently, jot it down in a notebook or on your phone, or share it with the person who has joined you on your walk.

6. **Find a gratitude partner.** Developing a new habit is usually easier with the support of someone else. Find a gratitude partner who can join you in the process. Try enrolling your spouse, partner, or children into a gratitude exercise conducted during dinner. (My oldest child and I have been sharing one thing we're grateful for as part of our bedtime routine since he was in preschool.) Or institute a daily gratitude text exchange with your best friend.

In Real Life: Sebastian's Audition

Sebastian started playing the violin at age five and had been dreaming of touring in a symphony ever since. After his recent graduation from a prestigious music school, he struggled with orchestral auditions because of his performance anxiety. During his auditions, he worried about making a mistake and got distracted scanning the committee's faces for their reactions. His biggest difficulty was handling the physical symptoms of his anxiety. His heart raced, his hands would shake, and he experienced a tightness in his chest, which wreaked havoc on his concentration and ability to play well. When Sebastian started therapy, two friends from music school had just won positions with first-tier orchestras, exacerbating Sebastian's stress and sense of urgency.

Strategy: Progressive Muscle Relaxation

With his most important orchestral audition coming up in two months, Sebastian practiced progressive muscle relaxation. I guided him to relax his mind and body by progressively tensing and relaxing muscle groups. We first practiced in session, then he continued on his own by listening to a recorded relaxation script. Every night before bed, Sebastian would play the 15-minute script and conduct the exercise. After a couple of weeks, he learned the script and performed the exercise from memory.

The Outcome

On the morning of his big audition, Sebastian woke up early and took his time doing the progressive muscle relaxation. He felt more relaxed and at ease. He arrived early to the audition and found a warm-up room to practice. When he was called to perform, he could feel his performance anxiety creeping in, starting with his shaky hands and tight chest. This time, he wasn't scared by these sensations, and was able to relax the areas of his body that were becoming tense. He played better than he had at previous auditions and felt more confident. A few months later, at his next audition, after trying out twice Sebastian finally won the role of second violinist in a world-class orchestra. He was ecstatic. Sebastian continues to practice progressive muscle relaxation every night and has shared how much this has positively contributed to his playing as well as his confidence.

GRATITUDE INVENTORY

The simple habit of keeping a list or inventory of things that you're grateful for can grow your happiness and gratitude over time. You get to experience the benefits of gratitude in the moment of writing it down as well as when you look back on your list in the future.

You can structure your inventory with different categories: people, experiences, places, animals, health. You can write everything in one sitting or give yourself weeks and months to keep adding to it.

Keep your list somewhere visible, such as on your refrigerator or pinned to a bulletin board next to your desk, so you are often reminded of all the things for which you are thankful.

You might also consider sharing your gratitude list with a friend or family member, particularly if they are part of it.

How to do it:

Use these 21 prompts to start your gratitude inventory in your notebook, and expand your list as you see fit.

Family members for whom I'm grateful are:
Because:

One thing about myself for which I'm grateful is:
Because:

Friends for whom I'm grateful are:
Because:

Something in nature for which I'm grateful is:
Because:

A challenge for which I'm grateful is:
Because:

GRATITUDE INVENTORY (continued)

A trip for which I'm grateful is:

Because:

A tradition for which I'm grateful is:

Because:

A teacher for whom I'm grateful is:

Because:

A strength of mine for which I'm grateful is:

Because:

A memory for which I'm grateful is:

Because:

An object for which I'm grateful is:

Because:

A life experience for which I'm grateful is:

Because:

A once-in-a-lifetime opportunity for which I'm grateful is:

Because:

An aspect of my work for which I'm grateful is:
Because:

An act of kindness done for me for which I'm grateful is:
Because:

A lesson for which I'm grateful is:
Because:

A gift for which I'm grateful is:
Because:

An animal for which I'm grateful is:
Because:

A mentor for whom I'm grateful is:
Because:

A talent or ability for which I am grateful is:
Because:

A childhood hero for whom I'm grateful is:
Because:

CHAPTER SIX

Build Communication Skills, Boost Self-Confidence

Learning new communication skills can be life-changing. This was true for one of my clients, Vivek, who was stuck in old patterns of not being able to say no, letting things slide for fear of causing conflict, and doing anything it took to make sure he didn't disappoint people. He frequently ended up resentful, upset, and frustrated in his personal and professional relationships. He was living a disempowered life and was tired of not feeling free, self-expressed, nor powerful.

Communication skills empower us to be in action, *creating* connection with others, instead of waiting or hoping for it to happen *to* us. Vivek learned how to listen to others in new ways that left people with the experience of feeling heard, seen, and understood. By doing so, he *generated* affinity, harmony, and connection. He also learned how to communicate in ways that gave him opportunities to experience the joy and freedom that came from being himself, and authentically connecting with other human beings from this place.

In this chapter, we'll look at communication skills you can learn and practice to create connection and ease in your interactions with others. What's more important than the tools, though, are your courage and commitment to bring connection *to* your life instead of trying to seek it out from someone or a situation. By approaching communication from that stance, you are already living in a way that's powerful and effective.

Communication Skills: A Primer

Now that you have the tools for handling negative thoughts and feelings that keep you out of social situations, it's time to add some practical skills to make your social encounters as successful as possible. Strong communication skills will not only help you build rapport, develop relationships, and grow your support network, you'll also enjoy greater self-confidence when interacting with other people. Studies have found that enhancing your communication skills reduces social anxiety symptoms, so these techniques do double duty: They enable you to convey your feelings, thoughts, and ideas clearly, and they help you continue to triumph over social anxiety. They can be used in combination with exercises throughout this book. For example, you can use conversation skills to start a chat with a stranger as part of an exposure hierarchy or remind yourself of your public speaking skills to replace negative thoughts about making an upcoming presentation.

As with all skills, the key to improving your communication skills is practice, practice, practice. Happily, you'll have plenty of opportunities. Whether you're chatting with someone at the checkout counter, sharing an anecdote with a friend, or expressing disagreement with a colleague at work, almost every day offers chances to exercise your social muscles.

Of course, practicing something new and outside of your comfort zone inevitably means you'll fumble sometimes, make mistakes, and have awkward interactions. This is part of any learning process, and it doesn't mean a social encounter was a failure. Mistakes show that you're being brave, you're on the court, and you're learning how you can improve so you can give up being hard on yourself. Your confidence will grow with more practice.

And remember, the goal isn't to pressure yourself to impress everyone you meet. Sometimes our expectations can get in the way, like assuming every chat is going to go perfectly or lead to a new friendship. As I like to remind my clients: You don't have to be *interesting* in every conversation, just *interested.*

Here are the skill types we're going to cover in this chapter:

Listening skills. Because we focus so much on finding the perfect words to say, it's easy to forget that listening is the foundation for great communication. With strong active listening skills, you'll stay engaged in conversations and help the other person feel heard and understood.

Conversation skills. Of course, eventually it's your turn to speak. Good conversation skills enable you to connect with another person and converse effectively. This includes knowing how to join a conversation and how to bail out gracefully when it's time to move on.

Nonverbal communication skills. Speech is only one way to communicate; nonverbal signals like body language, facial expressions, gestures, posture, and eye contact all play a role.

Interview skills. Nearly everyone finds job interviews to be intimidating. Techniques for managing this specialized social interaction will help you prepare for every aspect of the meeting.

Assertiveness skills. It might be uncomfortable to advocate for your own needs, but with practice, you can get used to speaking up for yourself. Learning how to communicate assertively allows you to transform potential conflicts into effective conversations. We'll explore nonviolent communication, an effective style of assertiveness.

Dating skills. How do you start and maintain a one-on-one conversation with someone you don't know? How do you listen attentively, share yourself authentically, and assert boundaries? Dating skills allow a more comfortable experience so you can relax and enjoy yourself.

Public speaking and presentation skills. Whether it's a presentation at work or a speech at your best friend's wedding, these skills help you deliver your message clearly so others can better understand your ideas and insights.

Interpersonal Communication Skills

Effective communication is more than a method for navigating social encounters; it's a vital life skill that can help you deepen your relationships, manage conflict, solve problems, and build trust with others. If social anxiety has been keeping you from opportunities to work on communication skills, don't worry, it's never too late.

Keep these points in mind as you try out your new skills:

There's always another chance. Opportunities to communicate abound, so don't pressure yourself to become a masterful communicator immediately. Every social interaction is a chance to get a little bit better. You could even practice with the next conversation you have today.

Communication goes both ways. A social interaction is an active, two-way process that involves both sending and receiving information, speaking and listening, and a shared understanding of what is being communicated. Be sure to work on both expressing yourself clearly and picking up on what the other party is trying to say.

Don't overthink it! If you're planning out everything that you're going to say next so that it sounds perfect, you may be too distracted to follow a conversation. Overly focusing on the other person's nonverbal communication, especially facial expressions, also gets in the way of the natural flow. Remember your mindfulness techniques so you can stay in the moment and use your communication skills naturally.

Listening Skills

Active listening is one of the most powerful ways to communicate and to cultivate relationships, whether it's a personal or professional connection. Unlike silently waiting for someone to finish speaking so you can reply, active listening involves paying full attention, asking questions, and reflecting back what someone is saying in your own words. The goal of active listening is for the other person to feel that they're being heard and understood; when a speaker senses that you're interested, and trying your best to understand, the result is more openness and honesty in the interaction. This naturally builds an atmosphere of trust and grows relationships, and over time often transforms an acquaintanceship into a friendship.

Another benefit of active listening is that it reduces misunderstandings. When you're actively listening, you're assuring the speaker that you're receiving the message as it was intended—or offering an opportunity for a misinterpretation to be corrected. This prevents misunderstandings that could potentially hurt a personal or professional relationship. Try these methods to use active listening in your conversations.

Give running feedback. Giving verbal and nonverbal responses while you listen will show the speaker that they have your attention. You might smile, nod your head, or give eye contact. Offer brief verbal affirmations: "I see," "I understand," or "Yes, that makes sense." When appropriate, offer empathetic responses like "I'm sorry you have to deal with this" or "That sounds tough."

Listen, don't advise. Withholding advice, unless you're specifically asked, is the best practice. When sharing a problem, most people prefer acknowledgment or empathy, and the space to figure things out on their own, instead of listening to someone's well-intended advice. The purpose of active listening is to increase understanding, not to problem-solve.

Ask, paraphrase, summarize. This trio of important responses will show that you're listening and make sure you're not losing the thread: ask clarifying questions, make paraphrasing statements, and summarize what you've heard. Clarifying questions can clear up any confusion and start with something like "Did you mean that . . ." or "Tell me more about . . ." When you paraphrase, you express what you've heard back to the speaker in your own words, showing that you're trying to gain a deeper understanding. Paraphrasing responses might start with "It sounds like . . ." or "What I heard you say was . . ." A summarizing statement is similar but pulls together several key points of the discussion: "Let me summarize what I've heard so far . . ."

SIX SCRIPTS FOR ACTIVE LISTENING

Active listening is a valuable technique that requires us to give our full attention, respond, and reflect on what is being said in a conversation. Here are the effective strategies for successful active listening, along with some example dialogue you can have at the ready for your next social interaction.

1. **Give Feedback:** Show the speaker you're listening with short responses.

   "I see."

   "Yes, that makes sense."

   "Please continue."

   "I'm listening."

   "That sounds tough."

   "I'm sorry you have to deal with this."

2. **Clarify:** Ask questions to confirm what the speaker has said.

   "Did you mean that..."

   "Tell me more about..."

   "Let me see if I'm clear..."

   "Could you explain to me again..."

   "I'm still not sure I understand..."

3. **Reflect:** Relay that you understood the speaker's feelings.

   "Sounds like you're feeling pretty frustrated and annoyed."

   "You seem upset and disappointed."

   "So you feel relieved?"

   "I get the sense that you might be feeling afraid about..."

   "I'm wondering if you're feeling hurt?"

4. **Paraphrase:** Reiterate what you think you heard.

   "It sounds like..."

   "What I heard you say was..."

   "You said...Do I have that right?"

   "If I'm hearing you correctly..."

5. **Summarize:** Relay the key ideas and themes the speaker has expressed.

   "What I've heard so far is..."

   "So the main problems you have with this are..."

   "It sounds like...is really important to you."

   "These seem to be the key ideas you've expressed..."

6. **Validate:** Appreciate the speaker's efforts.

   "I'm glad we're talking about this."

   "It makes me feel good that you confided in me."

   "I appreciate your willingness to talk about such a hard issue."

Nonverbal Communication Skills

We all engage in nonverbal communication, also known as body language, all the time. Our facial expressions, eye contact, gestures, posture, tone of voice, and body movements can express what we're feeling, sometimes more than words alone do.

Social anxiety can cause you to unknowingly signal disinterest or distance via "closed" body language. Closed posture involves hiding the trunk of your body by crossing your arms and legs. It can convey unfriendliness and anxiety. With open posture, we avoid blocking our trunk; this conveys openness and friendliness. Lack of eye contact, looking down, crossing your arms in front of you, or facing slightly away from someone might help you deal with anxiety during a conversation. But it can also make it appear you're not interested in communicating.

Of course, you now have other methods—CBT techniques, ACT skills, mindfulness practices—to cope with anxiety while conversing. A few adjustments can change the body language habits that social anxiety may have given you.

Go for 50 percent eye contact. The most important nonverbal cue to work on is looking into people's eyes, which is so important for maintaining the flow of conversation. If you're not used to this, a rule of thumb is to maintain eye contact about half the time, for about five seconds. This shows that you're interested, but it won't seem like you're staring. When you're starting to build this skill, you'll probably find it easier to hold eye contact while the other person is speaking and work up to maintaining it while you talk.

Adjust your posture. Leaning toward the speaker a bit is an easy shift that shows you're paying attention. It also helps to hold your head up, with your face toward the other person. Smile when it's appropriate, and let your hands rest in your lap or at your sides rather than crossing them or holding them in front.

Work on one at a time. Though these aren't difficult skills to hone, it can be helpful to pick one skill to improve at a time. You can also practice them in stages. For example, you could practice looking in your own eyes in the mirror first, then at strangers' eyes while walking down the street, then during brief interactions at a checkout counter, and then try maintaining eye contact with an individual or a group in conversation.

Conversation Skills

The art of conversation can sometimes feel confusing and frustrating for someone dealing with social anxiety. Just remember that you don't have to sound intelligent or try to impress everyone you meet. What's most important for effective conversations is being interested and curious. Making mistakes in social situations happens frequently to all of us. It's normal and part of being human.

Have a go-to opener. Learning good practices for starting and ending conversations is a great way to build confidence. Avoid a yes-or-no question as your conversation opener; try something that will evoke a more open-ended answer: "What are your plans for this weekend?" or "How do you know so-and-so?" or "Are you enjoying this weather as much as I am?" (Talking about the weather may be a clichéd icebreaker, but there's a reason: It's an easy, noncontroversial topic that anyone can weigh in on, giving you both a chance to get comfortable with each other.)

Joining a conversation that's already taking place can be intimidating, but most people know what that's like and won't begrudge you for it. An easy strategy is to listen attentively and wait. Either someone will draw you into the discussion, or you can add a comment, ask a question, or introduce yourself during a pause. Or throw one of your opening lines to the group—you'll have time to catch your breath while everyone responds.

Prepare an exit strategy. When you're feeling ready to end a conversation, often the easiest approach is to be direct. For example, if you're at a networking event, you can say, "It was a pleasure talking with you. Thanks for sharing your thoughts. Enjoy the rest of your evening." You can also ask the person for an introduction, such as "I promised myself I would meet three people. Is there anyone else you recommend I talk to?"

At most social gatherings, it's understood and expected that guests will move around and talk with different people, so you shouldn't feel bad about moving on. (Likewise, don't take it personally if someone does it to you.) You can say, "It was great talking with you. I'm going to look for my friend(s)," or "I'm tired and going to call it a night." And you can always end a conversation by asking for directions to the restroom. Just make sure you don't walk away in a different direction! If you're ending a conversation with someone you know, you can say, "I'll let you get back to (your work, your dinner, the rest of your day . . .)" That's a good exit for phone conversations, too.

Follow the three-to-one rule. Many people who are socially anxious find small talk to be stressful, because they worry about how to respond or fear saying something wrong. These casual conversations are worth having, however; they can be the springboard for transforming discussions into relationships. Just remember that with small talk, the focus is not on you, it's on the other person. Richard S. Gallagher, author of *Stress-Free Small Talk*, invented the "three-to-one" rule: For every three questions you ask the other person, share something about yourself. A four-to-one or five-to-one ratio is fine, too, if that's more comfortable. Gallagher notes that the best kind of questions to ask will speak to people's joy and strengths. So try to find out what the person feels proud of, excited about, or thinks is important.

Openers and Closers

For easy reference, here are 10 great all-purpose lines for starting and ending conversations:

Conversation starters:

1. What are your plans for the weekend?
2. What a beautiful/rainy/cloudy/sunny day.
3. What brings you to (this setting)?
4. How are you connected to (host)?
5. What do you like to do when you're not working?

Conversation enders:

1. It was great to talk with you.
2. Is there anyone else I should be sure to meet?
3. Can you direct me to the restroom?
4. I'll let you get back to . . .
5. I'm going to go find my friends.

Interview Skills

Interviewing can be intimidating and even daunting. Some of the pressure results from thinking you have to come up with flawless responses. But whatever industry you're in, the goal is for the interviewer to determine if you're a good candidate for the position and an asset to the company or organization. You can be very qualified for a job but struggle to show this. When you're grappling with social anxiety, the idea of being judged under a spotlight by people you don't know—or worse, a panel—can be especially anxiety-provoking.

Here are some tactics for a successful interview:

Prepare. Learn everything you can about the company and the position. Take time to consider how your experience and qualifications make you a great fit for the job and the employer's goals. This not only enables you to address it during the interview, it will boost your confidence knowing that you've done your homework.

Practice. Draft as many questions as you can that an interviewer might ask you. Interview questions commonly relate to your professional strengths, weaknesses, accomplishments, and any challenges you've faced. Why do you want to leave your current job? Why do you think you're a good fit for the role and the company? Where do you see yourself in 5 or 10 years; what are your goals for the future? An interview is not about answering all questions correctly; the interviewer wants to understand who you are and how you can help the company. Use your list of questions to stage practice interviews with friends or family.

Be prepared to ask questions. Inevitably, the interviewer will give you an opportunity to ask questions of your own, so be ready. Avoid questions that you could easily have answered by looking at the employer's website or doing basic research. Remember your conversational skills and ask open-ended questions about the position and how it fits with the company's goals. The main point is to appear interested in succeeding at the position you're interviewing for.

Send the right message. Make sure your nonverbal cues send the right signals. Leaning in and sitting up straight shows that you're interested. Avoid crossing your arms. Make sufficient eye contact and use your active listening skills to show you care.

Follow up. After the interview, it's a nice gesture to send a thank-you note via email to your interviewer or hiring manager. It shows politeness, demonstrates conscientiousness, and keeps you in their minds. If you don't get the job, look back on the interview experience and consider what you can improve on for your next interview.

Assertiveness Skills

People with social anxiety tend to avoid conflict, so you might struggle with being assertive. Sometimes being honest with others about something you need can feel risky and make you feel vulnerable. But assertive communication honors that you have the right to stand up for what's important to you. Being able to say no when you need to, or ask for what you need, eliminates resentment, frustration, stress, and feelings of helplessness.

Nonviolent Communication, or NVC, is a simple yet highly effective communication tool first developed by psychologist Marshall Rosenberg, PhD, in the 1960s. NVC guides us to share what is being observed, felt, and needed in any given situation. Its principles offer a method for making requests of others in a way that leads to harmony, connection, and a mutually beneficial outcome.

NVC has four basic components that you can use to create a statement that asserts your need: observations, feelings, needs/values, and requests (OFNR).

How to do it:

Here's what an OFNR statement looks like:

Observation: In your notebook, share what you objectively see or hear, without blame. Imagine what a video camera would observe to make sure you're not adding evaluation.

When I hear/see . . .

Feeling: Express your emotions.

I feel . . .

Need/Values: Say what you are wanting or needing, or what's important to you.

Because I need/value . . .

Request: Make a specific request to help fulfill your needs.

Would you be willing to . . . ?

Example 1:

Observation: When I hear you say you won't have the presentation ready for our clients until Friday . . .

Feeling: I feel concerned . . .

Need/Value: Because I value reliability and efficiency.

Request: Would you be willing to tell me what's getting in the way of completing the report and how I can help you complete it by the deadline?

Example 2:

Observation: When I see clothes on the floor of the living room . . .

Feeling: I feel irritated and upset . . .

Need/value: Because I value order in our home.

Request: Would you be willing to put your clothes in the laundry hamper or in your room?

More examples:

When I see you reading a magazine while I'm talking, I feel frustrated because I'm wanting to be heard. Would you be willing to put down the magazine for a few minutes and listen to some ideas for how we can spend the weekend?

When I heard you say that you can't come over tonight like we'd planned, I feel sad and disappointed because I want to be connected. Would you be willing to talk on the phone tonight instead?

When I notice you've arrived 20 minutes late, I feel a little annoyed because I value consideration and dependability. Would you be willing to let me know in advance when you're not able to meet at the time we agreed on?

Your turn: Read each scenario below and craft an OFNR statement to address it.

Scenario 1: Your roommate is playing music loudly while you're trying to sleep.

Scenario 2: Your sister wants you to come over and help her pack for her trip, but you had planned to spend your evening relaxing at home after a hard day.

Scenario 3: Your friend asks if you can pick her up from the airport on a Friday during rush hour, and you've had a long week and are looking forward to going home and relaxing.

Dating Skills

Dating skills can help you find a partner, but they can also teach you how to be your true self and create genuine, natural connections with other people.

Dating can be tough. The anticipation of saying the wrong thing or coming across awkwardly can be all-consuming. Social anxiety can take a toll on your ability to initiate and maintain romantic relationships. It can feel hard to be vulnerable and let your guard down with someone you've just met or don't know well yet. Rest assured that from asking someone out to ending a date, dating doesn't have to feel as nerve-wracking.

Make a specific ask. When asking someone out, either online or in real life, it's helpful to be specific: "Would you like to go out for coffee or sushi dinner?" rather than "What are you weekend plans?" Give multiple options, like suggesting Friday or Saturday, or asking the person where they like to eat. Sometimes it's more comfortable for both parties to commit to a date that's short in duration, like a 30-minute coffee date at a café.

Be willing to share. While you're on a date, if you're feeling nervous, you can use Gallagher's three-to-one rule and ask three questions before you disclose something personal about yourself. Be sure that you do disclose, though. Intimacy is based on understanding each other. It's hard to create a connection if you don't share things about yourself.

Don't practice! Unlike a job interview, where you should always rehearse and prepare your answers, it's best not to prepare what to say on a date. Trying to recall a predetermined script can backfire and make you appear more nervous. Talk about what's important to you, use your active listening and conversation skills, and enjoy the spontaneity—even the awkward moments—of getting to know someone.

Remember your exit strategy. Ending a date is similar to ending a conversation; being gracious and polite goes a long way. If it doesn't feel like a match, you can end by saying something like "I should be going now. Thanks for a nice evening" or "Thanks for your time." If the date went well, you can share that you enjoyed your time together and would like to do it again. If your date gets in touch with you after, but you're not interested, you can say, "I didn't feel the connection I'm looking for, but I wish you the best."

Follow up with a plan. If things went well, don't get hung up on how long you should wait before reaching out. Follow your intuition, based on your feelings for the other person and what you learned about them. What's more important is that your follow-up includes another specific plan: "What are you up to on Friday? Would you like to go out for Thai food with me?"

In Real Life: Desmond's Party Conversations

Desmond was disheartened when he first came to see me. He had always felt shy and awkward approaching people to start conversations. He either didn't know what to say or viewed small talk as forced and stilted. Desmond had a few close friends but wanted to expand his circle. He knew it would be good to at least meet some friends of friends, but he'd been avoiding parties for years. When he forced himself to go to a friend's birthday party a few months prior, the experience left him feeling confused. He didn't know what to do with himself, how much he should mingle, or how to break into a group conversation. He ended up spending his time talking to the friend he came with, then left after a half hour.

Strategy: Improving Conversation Skills

Desmond and I worked on conversation skills together. First, he recognized that he had unreasonable expectations for himself. He thought he needed to be entertaining, be engaging, and have great conversations with everyone. He now understood that he wasn't going to hit it off perfectly with everyone, and that being a good listener is an equally important part of socializing. We practiced new ways to start and end conversations as well as how to ask open-ended questions and disclose information to keep the flow going. He understood that it's okay to join a group and listen before making a comment or introducing himself. We did role-plays in session, and as homework, he practiced having conversations with at least one new person each day.

The Outcome

At the next party, Desmond was ready to mingle. He promised himself he would start a conversation with at least three new people. He walked up to the first person who looked friendly and asked how she was connected to the host, which led to some standard get-to-know-you questions and answers. Desmond talked to a few more people and was able to keep the conversations going with open-ended questions and by sharing a little about himself. One conversation fell flat, so he excused himself to the restroom. Finally, he saw a small group talking in the kitchen and mustered the courage to join in. He hung back for a few minutes and listened attentively. When there was a pause, he introduced himself and asked how everyone knew each other. Desmond was pleased to discover they had attended the same university as he had. Desmond came to therapy the next week feeling satisfied and motivated to try his skills again.

Performance Skills

If you dread the thought of speaking in front of a group, you are not alone. Most people feel nervous about giving a speech or presentation, even seasoned experts. When social anxiety is involved, public speaking goes from a common dread to a worst nightmare. As we review some helpful pointers to ensure your success, keep these points in mind.

The audience isn't looking for perfection. Typically, the most uncomfortable part of public speaking is trying to appear as flawless and brilliant as possible and using this as your measure of success. A key thing to remember, whether presenting to a small group, giving a wedding toast, or making a keynote speech, is that the majority of the people in the audience genuinely want to see you succeed. They aren't secretly waiting for you to fail so they can laugh. In fact, since most people fear public speaking to some extent, they tend to empathize quite easily with a speaker who seems a little nervous. Think about the last time you saw someone present or speak in public. My guess is that you weren't hoping they would fail so you could look down on them, right?

You're there to share. The real goal of most presentations is to teach people something. If you share even a single nugget of wisdom with your audience—even if it's a heartwarming anecdote about how the groom met the bride—you've offered something of value. Shifting your view of success in this way can take some of the pressure off. I've had many clients shift from feeling panicky to being genuinely excited by the thrill they now experience from a chance to impart their ideas.

The experience has its rewards. Developing good public speaking skills can have an impact in many areas of your life. Being an effective presenter and public speaker not only improves your confidence, it opens up more opportunities in your career.

On a deeper level, improving your public speaking can enable you to influence the environment around you. The power of speech can be a force for positive change.

Getting the Audience Interested

When you speak, it's crucial to engage your audience. Not only does this keep people interested in your message, it can also help you feel less isolated as the speaker.

Make a connection. A good move to start things off right is to make eye contact with the most welcoming faces in the audience. As you introduce yourself, also consider saying something that alludes to the previous speaker's message or theme, if appropriate, to connect with what the audience is feeling.

Change things up. It's good to remember that people can find it hard to pay attention to something after 15 minutes. For this reason, building frequent minor changes into your performance, such as showing a video clip or telling a story, will help sustain the audience's interest and attention, particularly during a longer presentation.

Keep the momentum going throughout the presentation. Designating times throughout your talk to answer questions is more interesting than saving questions for the end. Also look for chances to quiz your audience, asking them to raise their hands for a true/false or "How many of you have ever . . ." question.

SIX KEY WAYS TO ENGAGE YOUR AUDIENCE

Listening to a formal talk for any length of time is challenging for any adult. When your audience disengages, they fail to absorb your ideas. The following six strategies outline the important elements for getting your audience engaged with your talk so they can better hear and digest the ideas you're presenting.

1. **Tell a story.** People listen attentively when you're telling a story, because they want to know what happens next.The best story is one with a conflict that needs to be resolved. If you can't create an overall narrative for your whole presentation, look for opportunities to use short anecdotes to make your points.
2. **Add video clips.** If this is an option, including video clips in your presentation can break the monotony, reinforce your message, and evoke feelings that are more difficult to elicit with just speaking. Videos engage the senses, explain a complex idea more easily, and establish an emotional connection with the audience. Keep video clips short and simple.
3. **Ask for questions.** An adult's attention wanes after about 10 to 15 minutes. Taking a break from presenting to interact with your audience helps regain their attention. Instead of waiting until the end or after, you can break to ask for audience questions throughout your presentation.
4. **Survey the audience.** Polls are another tool for allowing your audience to take a mental break from listening to you speak. This can be as simple as asking the audience a series of questions, starting with: "Raise your hand if..." Alternatively, you can use polling software and have your audience answer with their smartphones, and have the results appear on your screen.
5. **Try paired sharing.** An interesting way to raise engagement is to have audience members talk to one another. You can ask the audience to pair up with the person next to them and give them a topic to discuss or a technique to practice. You can let them take turns and remind them when to switch. After a designated time, continue your presentation by polling the audience on the results of their discussion.
6. **Encourage movement.** Having your audience move their bodies is an interactive way to shake off the fatigue of sitting and raise the energy level in the room. This can be as simple as asking everyone to stand up and stretch for a few seconds.

Organizing Your Speech

A well-organized speech can increase your confidence in presenting ideas. If your presentation is not structured thoughtfully, your audience will have to search for the message, which is hard work. A helpful structure is to introduce your main idea, then follow it up with key points.

Consider the audience. Think about what's of most interest to the group you're speaking to. What is their main concern? How can you make that front and center in your presentation?

Start with an outline. For the audience to know what to expect, you can outline your message and list the main points you'll be touching. This will also help you stick to those points.

Close with emphasis. At the end of your presentation, it's helpful to repeat the most important points that you've made to make sure everyone takes in the message you're imparting. If you can include potential future outcomes, a sense of what could be, as well as what action steps are needed to get there, you'll close with your audience inspired.

Visual Aids

We learn best through our senses. Audiovisual aids—video, maps, graphs, charts, photographs, illustrations—add variety that engages the audience and can reinforce your message and key points.

It's helpful to limit each visual element to one topic and keep images as simple as possible so they can be grasped quickly and seen easily from a distance. Be consistent in the color and design elements throughout your presentation. Remember that the audience won't be able to focus on the images you're showing and listen to you talk at the same time. Build in pauses for them to take in what they're seeing before you continue.

Practice

Practicing your public speaking skills can alleviate anxiety as well as help you polish your presentation. You can start by practicing in front of a mirror or by recording yourself with your phone so you'll get used to seeing and hearing yourself as you refine and revise. Text that seems effective on the page may be difficult or awkward to speak, so always recite your script out loud to see how well it works. Time yourself while you rehearse to gauge if your speech or presentation is under or over time.

Doing a practice run with friends, colleagues, or family members will help you feel more at ease. Even more useful is to practice with someone who will give you honest feedback. Instead of asking if they liked it, you can ask what element they liked the best and why, or what they would like to see improved for next time.

You'll want to rehearse multiple times, even if you're not memorizing the text. When you're nervous, it's easy to talk too fast during your presentation. When you're very familiar with presentation, you'll be more relaxed. Slowing down and allowing for pauses in your speech will help your talk feel more conversational and get your message across more powerfully.

Breathing and Speaking

Anxiety before or during a speech or presentation makes us breathe more quickly and shallowly. This not only impacts the pitch and tone of our voice, but the sensation of gasping for air can make us more nervous or even panicked.

Breathing for speech requires diaphragmatic breathing, also called "belly breathing," which helps deepen your breath so you're not just using the upper part of your lungs to get air. You can practice this by placing one hand on your abdomen and feeling how it moves in and out as you breathe slowly.

Standing tall is the best posture for working with your breath while speaking. When you get in the habit of diaphragmatic breathing like this, you'll feel calmer and you'll come across more poised when you speak. Even if you practice for a few minutes a day, the results are cumulative.

Here's a more detailed breathing exercise.

DIAPHRAGMATIC BREATHING FOR VOICE QUALITY

Your voice is an important tool when you're speaking in front of others. Breathing properly can help you achieve a stronger voice with richer tone and better projection.

When you're not breathing deeply enough, you might speak at a low volume or have a breathy voice and wavering pitch that is harder for the audience to hear and understand. When your breath is shallow and coming from just your chest, you may not have enough air to finish your sentences.

By improving the quality of your voice through diaphragmatic breathing, like singers do, you can feel more confident and comfortable on stage.

How to do it:

1. Lie on the floor.
2. Place your hand on your abdomen.
3. As you breathe in, feel your abdomen rise, like a balloon inflating.
4. As you breathe out, feel your abdomen contract, like a balloon deflating.
5. Inhale to the count of 8 and exhale to the count of 8.
6. Repeat three times.

After you've become familiar with it, you can also practice this breathing exercise while standing. Diaphragm breathing may feel awkward at first, because most of us are used to breathing from our chest. Practice as often as you can, especially leading up to your speech or presentation, and it will become second nature.

In Real Life: Lily's Presentation

Lily had been passing up promotions at work for the last five years due to her performance anxiety. Each time she turned one down, she rationalized that any position that requires giving a presentation wouldn't be worth the weeks—or worse, months—of anticipatory stress. Her last presentation was in graduate school when she was supposed to present her research findings. For weeks, she agonized over her slides. During her presentation, she was so out of breath that she could hardly speak, and she felt like she was on the verge of a panic attack. The worst part was when she looked out at the audience near the end of her presentation. She was mortified to see that nearly a quarter of her fellow students had snuck out the back of the room, and the other half were looking at their phones.

Strategy: Improving Performance Skills

Lily was offered a promotion with a bonus she couldn't turn down. But it entailed making presentations once per quarter. After taking the position, she started learning about how to give effective presentations. She studied ways to engage the audience, like stopping to ask questions, showing video clips, or sharing a story. Lily took an online course on designing audiovisual aids and made sure her slides were concise, used color carefully, and used images that emphasized her message. She practiced with a few close friends in the weeks leading up to her first presentation and incorporated their suggestions. Lily practiced breathing exercises to stay relaxed and enhance her speaking voice.

The Outcome

As soon as Lily got to work on the day of her presentation, she rehearsed one more time, reviewed her cue cards, and took one last look at her slides. She could feel her jitters increasing as the minutes went by but reminded herself that everyone gets nervous and that she had tools to manage it this time. She spent the remaining minutes before people arrived practicing diaphragmatic breathing. With one hand on her belly, she felt her diaphragm move up and down as she took slow, deep breaths. When it was time to start her presentation, she felt calm, confident, and ready. Her presentation was a success and she felt a huge sense of relief.

Conclusion

Congratulations! This is a perfect moment to pause and acknowledge all of the progress you've made. It takes willingness, time, effort, and bravery to learn and try out new strategies, so be sure to give yourself credit and appreciation.

What's important now is to stay courageous and powerful in the face of the inevitable and natural challenges and setbacks that will come. Staying on track by actively using everything you've learned will ensure you're not left with just insights and good ideas but can continue to fulfill your goals and produce permanent, lasting results.

In this final chapter, we'll review your victories and challenges, look at where and why you might be stuck, and how to stay the course. We'll explore the best ways to set yourself up for success on the rest of your journey.

What Have You Learned?

We've come a long way! This book has presented you with a myriad of essential tools for shifting your social anxiety. Let's review the essential strategies so you easily recall what you've learned and can implement them in the areas of your life that matter to you.

What is social anxiety? We began with an in-depth look at the mental and physical symptoms of social anxiety and the social situations that can trigger your fears. We looked at how our brain's threat detection system can go haywire, allowing social encounters to hijack our attention and setting off our body's stress response.

Changing your thoughts. We covered methods from cognitive behavioral therapy that can help you uncover and defeat the negative thoughts that are causing you distress, and replace them with realistic, positive ones.

Facing your fears. We explored how exposure techniques, like graded exposure and systematic desensitization, can interrupt your pattern of avoiding anxiety so you can safely face your fears and experience long-term relief.

Accepting and committing. Acceptance and commitment therapy showed us how being present and willing to accept the discomfort of our thoughts, feelings, and bodily sensations can loosen anxiety's grip. Along with that, we saw that clarifying your values, setting goals, and taking action enables you to move in the direction of what your heart most desires.

The power of mindfulness. We examined techniques to cultivate calm and reduce stress through mindfulness and meditation. We learned that mindfulness can be practiced in everyday moments and through formal techniques like meditation. We explored how gratitude and self-compassion practices can increase joy and happiness.

Sharpened skills. Last, we looked at essential communication skills you can improve on, from active listening to public speaking skills, to handle social situations with more confidence and ease. These techniques are not only helpful from a practical sense; the confidence they give you helps diminish social anxiety.

I invite you to look back at your self-assessment and goal setting from chapter 1 to compare how far you've come. What benefits have you discovered from the work you've done? How has applying the strategies offered in these chapters helped you move closer toward your goals? Acknowledging and celebrating any progress you've made—from small wins to huge breakthroughs—enhances your confidence and motivation and is a great practice to cultivate.

Your Biggest Victories

What prompted you to pick up this book? How was social anxiety affecting you and where was it holding you back? What old habits were getting in the way of being confident and connected with others? What seemed impossible to overcome?

My hope is that you're thinking and acting in new ways. But ultimately, the measure of success is up to you. Maybe you're willing to face your fears and lean into discomfort. You may be saying yes to invitations and talking to yourself with compassion and reassurance when you feel nervous or want to back out. Or perhaps when interacting with others, you're trying out more small talk, eye contact, or active listening than you used to. Maybe you're more able to assert yourself so you can get your needs met. Only you can decide what victory looks like, and you have your values and goals to measure it by. The important thing is to acknowledge your successes as you keep growing your courage and confidence and trying out new practices and techniques.

Your Biggest Challenges

The idea of facing your fears can be uncomfortable and scary! You've most likely had some challenges and setbacks along the way. You may even be still hoping for a method to overcome your social anxiety that doesn't involve confronting what scares you. What have been your biggest challenges and setbacks? Have you gotten stuck anywhere along the way?

Obstacles are inevitable on any journey worth taking, but they don't have to stop you. You gain courage, strength, and confidence each time you get back up and take another step. When self-defeating or hopeless thoughts come up, like "I'm a failure" or "This won't work," you can remind yourself: "Setbacks are inevitable and a normal part of the course. They don't mean I'm a failure. Learning new ways of being takes time. I can try again tomorrow."

Being courageous and taking action are habits. Like any new habit, these take effort and intention to develop, especially in the beginning. Remember that, as you continue to practice, your new habits will eventually become effortless and second nature. You will build momentum and gain energy. Choosing courage over comfort is a choice that's always available to you, from one moment to the next.

Setting Yourself Up for Success

This book has presented you with a wealth of strategies to shift your social anxiety. Keep these final points in mind to set yourself up for success as you continue to move forward.

Actions create movement. The most important thing to remember is that actions—big or small—are what's needed to create results. Reading this book, then putting it back on the shelf won't do much to affect your anxiety. Even if you only make one small shift each day, like smiling at a stranger or saying good morning to a coworker, you will keep heading in a new direction.

Be honest with yourself. Creating new habits takes some effort, but it doesn't take long before you're naturally thinking and acting in new ways. Wherever you get stuck, I encourage you to honestly assess if you need to muster more courage, willingness, or participation in that area that's giving you trouble. If you've avoided any exercises, I suggest you go back and do them, starting with the ones you least want to do. Try to lean into what's unpleasant and participate in new ways that produce results.

Keep an eye on your values, goals, and actions. Remember to start your goal setting by connecting to what's most important to you in your life—your values. Then create goals in that area and plan out steps you can take today, within the next week, months, and year. Whenever you get overwhelmed, break the steps into smaller ones, and commit to doing at least one thing a day. As Annie Dillard, famous American author and poet, reminds us: "How we spend our days is, of course, how we spend our lives. What we do with this hour, and that one, is what we are doing." Hour by hour, we can choose new actions that improve the quality of our lives.

Building a Support Network

Despite your hard work and best efforts, you may find that overcoming social anxiety is hard to accomplish on your own. If you find you're struggling to start or complete the strategies, you're becoming overwhelmed, or you just need more accountability, this is a great time to build a network of support.

Working with a therapist can be highly effective. Receiving personalized guidance that's tailored to your challenges and needs can be life-changing. When you're searching for a therapist, I recommend looking for licensed mental health providers who specialize in CBT and/or ACT, exposure therapy, and mindfulness. As we've seen, these approaches have shown to be the most effective in treating social anxiety.

Support groups are another great resource that will broaden your network and reduce isolation. Connecting with people who deeply understand, relate, and share your struggles can help you feel more normal and less alone. Listening to other people's experiences can be inspiring and motivating. Joining a group can be a bold exposure exercise in itself; you're already winning just by signing up. Being part of a support group will immediately increase your sense of community and belonging.

Online organizations are also a source for support and resources. There are forums for people with social anxiety that you can join as well as online courses, blogs, and articles that can provide you with additional information and strategies.

Be sure to take a look at the resources section at the end of this book for help with all of the above.

If you're willing to be vulnerable and share your struggles, goals, and commitments with a friend or family member, you'll be unequivocally set up for success. Like a great workout buddy, a friend or loved one can help you stay accountable and even join you when it's helpful. When you experience setbacks, your supporters will be there to encourage you and remind you of your goals and values. When you make gains, you'll have people to celebrate with and cheer you on.

I'm so glad you've taken the time to read this book and try out strategies to transform your social anxiety. May you continue being courageous and leaning in when things are uncomfortable. Keep doing the work to generate a life you love.

Resources

Online Resources

Anxiety and Depression Association of America (ADAA)
https://adaa.org/
The Beck Institute for Cognitive Behavior Therapy
www.beckinstitute.org
Center for Nonviolent Communication
www.cnvc.org
Dr. David D. Burns TEAM-CBT
https://feelinggood.com
Greater Good Science Center
https://greatergood.berkeley.edu/
Guided Mindfulness Meditation practices with Jon Kabat-Zinn
https://www.mindfulnesscds.com/
Mayo Clinic
www.mayoclinic.org
Mindfulness-Based Cognitive Therapy (MBCT)
www.mbct.com
Mindfulness-Based Stress Reduction (MBSR programs)
https://mbsrtraining.com/
Mindful Self-Compassion (MSC)
https://self-compassion.org
National Institute of Mental Health (NIMH)
https://www.nimh.nih.gov/index.shtml
National Social Anxiety Center (NSAC)
http://nationalsocialanxietycenter.com
Social Anxiety Institute
https://socialanxietyinstitute.org
Very Well Mind
https://www.verywellmind.com

Finding Help

Academy of Cognitive Therapy
http://academyofct.org
Association for Behavioral and Cognitive Therapies (ACBT)
http://www.findcbt.org/
Association for Contextual Behavioral Science (ACBS)
http://contextualscience.org/act
Feeling Good Institute: TEAM-CBT
http://www.feelinggoodinstitute.com/
Good Therapy
http://goodtherapy.org
International Association for Cognitive Psychotherapy
http://the-iacp.com
Psychology Today
www.psychologytoday.com/us

Apps

Calm
www.calm.com
Headspace
www.headspace.com
Insight Timer
https://insighttimer.com

Books

Brach, Tara. *Radical Acceptance: Embracing Your Life with the Heart of a Buddha.* New York, NY: Random House, 2003.

Brown, Brené. *The Gifts of Imperfection: Let Go of Who You Think You're Supposed to Be and Embrace Who You Are.* Center City, MN: Hazelden Publishing, 2010.

Brown, Brené. *Daring Greatly: How the Courage to Be Vulnerable Transforms the Way We Live, Love, Parent, and Lead.* New York, NY: Penguin Random House, 2012.

Brown, Brené. *I Thought It Was Just Me (But It Isn't): Making the Journey from "What Will People Think?" to "I Am Enough."* New York, NY: Penguin Random House, 2007.

Burns, David D. *Feeling Good: The New Mood Therapy.* New York, NY: Penguin, 1981.

Burns, David D. *Intimate Connections.* New York, NY: Penguin, 1985.

Burns, David D. *The Feeling Good Handbook.* New York, NY: Penguin, 1999.

Burns, David D. *When Panic Attacks: The New, Drug-Free Anxiety Therapy That Can Change Your Life.* New York, NY: Random House, 2006.

Chödrön, Pema. *Comfortable with Uncertainty: 108 Teachings on Cultivating Fearlessness and Compassion.* Boston, MA: Shambala Publications. 2002.

Chödrön, Pema. *The Places That Scare You.* Boston, MA: Shambala Publications. 2001.

Fleming, Jan E., and Kocovksi, Nancy L. *The Mindfulness and Acceptance Workbook for Social Anxiety and Shyness: Using Acceptance and Commitment Therapy to Free Yourself from Fear and Reclaim Your Life.* Oakland, CA: New Harbinger Publications, 2013.

Hanson, Rick. *Buddha's Brain: The Practical Neuroscience of Happiness, Love, and Wisdom.* Oakland, CA: New Harbinger, 2009.

Hanh, Thich Nhat. *The Miracle of Mindfulness: An Introduction to the Practice of Meditation.* Boston, MA: Beacon, 1999.

Harris, Russ. *ACT Made Simple: An Easy-To-Read Primer on Acceptance and Commitment Therapy.* Oakland, CA: The New Harbinger Made Simple Series, 2009.

Harris, Russ. *The Confidence Gap: A Guide to Overcoming Fear and Self-Doubt.* Boulder, CO: Trumpeter, 2011.

Harris, Russ. *The Happiness Trap: How to Stop Struggling and Start Living.* Boston, MA: Trumpeter, 2008.

Hayes, Steven C., and Smith, Spencer. *Get Out of Your Mind and Into Your Life: The New Acceptance and Commitment Therapy.* Oakland, CA: New Harbinger, 2005.

Kabat-Zinn, Jon. *Wherever You Go, There You Are: Mindfulness Meditation in Everyday Life.* New York, NY: Hyperion, 1994.

Neff, Kristin. *Self-Compassion: The Proven Power of Being Kind to Yourself.* New York, NY: Harper Collins, 2011.

Neff, Kristin, and Germer, Christopher. *The Mindful Self-Compassion Workbook: A Proven Way to Accept Yourself, Build Inner Strength, and Thrive.* New York, NY: Guilford Press, 2018.

Salzberg, Sharon. *Lovingkindness: The Revolutionary Art of Happiness.* Shambhala, 2002.

References

American Psychiatric Association. *Diagnostic and Statistical Manual of Mental Disorders*, 5th Edition. Arlington, VA: American Psychiatric Publishing, 2013.

Anxiety Canada. "Effective Communication: Improving your Social Skills." www.AnxietyCanada.com, 2005.

Beck, Aaron T. *Cognitive Therapy and Emotional Disorders*. New York: International Universities Press, 1976.

Beck, Aaron T., Emery G. and Greenberg R. L. *Anxiety Disorders and Phobias: A Cognitive Perspective*. New York: Basic Books, 1985.

Beck, Judith S. *Cognitive Behavior Therapy: Basics and Beyond* (2nd ed.). New York, NY: Guilford Press, 2011.

Beck, Judith S. "Cognitive Therapy." *Corsini Encyclopedia of Psychology*. 1–3, 2010.

Beck Institute of Cognitive Behavioral Therapy. "History of Cognitive Behavior." Accessed February 20, 2020. https://beckinstitute.org/about-beck/team/our-history/history-of-cognitive-therapy/.

Bradt, Steve. "Wandering Mind Not a Happy Mind." *Harvard University Gazette*, November 11, 2010.

Bridges to Recovery. "Causes of Social Anxiety." Accessed January 12, 2020. https://www.bridgestorecovery.com/social-anxiety/causes-social-anxiety.

Brown, Brené. *The Gifts of Imperfection: Let Go of Who You Think You're Supposed to Be and Embrace Who You Are*. Center City, MN: Hazelden, 2010.

Boston University Office of the Ombuds. *Active Listening Handout*. Accessed March 29, 2020. http://www.bumc.bu.edu/facdev-medicine/files/2016/10/Active-Listening-Handout.pdf.

Burns, David D. *Feeling Good: The New Mood Therapy*. New York, NY: Penguin, 1981.

Burns, David D. *Intimate Connections*. New York, NY: Penguin, 1985.

Burns, David D. *The Feeling Good Handbook*. New York, NY: Penguin, 1999.

Burns, David D. *When Panic Attacks: The New, Drug-Free Anxiety Therapy That Can Change Your Life*. New York, NY: Random House, 2006.

Burns, David D. "Why Are Depression and Anxiety Correlated? A Test of Tripartite Model." *Journal of Consulting and Clinical Psychology*, 66, 461–473, 1998.

Carducci, Bernardo, and Zimbardo, Phillip G., "The Cost of Shyness: Shyness Is an Overgeneralized Response to Fear: and It's Easy to Beat Once You Understand This." *Psychology Today*, November 1, 1995.

Carliss, Julie. "Mindfulness Meditation May Ease Anxiety, Mental Stress." Harvard Health Publications. January 8, 2014.

Chen, Lung Hung, and Wu, Chia-Huei. "Gratitude Enhances Change in Athletes' Self-Esteem: The Moderating Role of trust in Coach." *Journal of Applied Sport Psychology*, 26(3): 349–362, May 2014.

Chödrön, Pema. *Taking the Leap*. Boulder, CO: Shambala, 2009.

Clark, David A., and Beck, Aaron T. *Cognitive Therapy of Anxiety Disorders: Science and Practice*. New York, NY: Guilford Press, 2010.

Cook, Gareth. "Why We Are Wired to Connect." *Scientific American*. October 22, 2013.

Cuncic, Arlin. "An Overview of Social Skills Training." *Very Well Mind*: December 9, 2019.

Cuncic, Arlin. "How to Socialize If You Have Social Anxiety Disorder." *Very Well Mind*: January 26, 2020.

Cuncic, Arlin. "10 Body Language Mistakes You Might Be Making." *Very Well Mind*, September 18, 2019.

Duarte, Nancy. *Slide:ology: The Art and Science of Creating Great Presentations*. Sebastopol, CA: O'Reilly Media, 2008.

Ellis, Albert. *Reason and Emotion in Psychotherapy*. Secaucus, NJ: Lyle Stuart, 1962.

Fleming, Jan E., and Kocovksi, Nancy L. *The Mindfulness and Acceptance Workbook for Social Anxiety and Shyness: Using Acceptance and Commitment Therapy to Free Yourself from Fear and Reclaim Your Life*. Oakland, CA: New Harbinger Publications, 2013.

Foa, E. B., Franklin, M. E., Perry, K. J., and Herbert, J. D. "Cognitive Biases in Generalized Social Phobia." *Journal of Abnormal Psychology*, 105: 433–439, 1996.

Gallagher, Richard S. *Stress-Free Small Talk: How to Master the Art of Conversation and Take Control of Your Social Anxiety*. Emeryville, CA: Rockridge Press, 2019.

Great Good Science Center. "What Is Gratitude? Why Practice It?" University of California, Berkeley. Accessed March 20, 2020. https://greatergood.berkeley.edu/topic/gratitude/definition#what-is-gratitude.

Hanh, Thich Nhat. *The Miracle of Mindfulness: An Introduction to the Practice of Meditation*. Boston, MA: Beacon, 1999.

Harris, Russ. *ACT Made Simple: An Easy-To-Read Primer on Acceptance and Commitment Therapy*. Oakland, CA: The New Harbinger Made Simple Series, 2009.

Harris, Russ. "The Complete Set of Client Handouts and Worksheets from ACT books."

Harris, Russ. *The Confidence Gap: A Guide to Overcoming Fear and Self-Doubt*. Boulder, CO: Trumpeter, 2011.

Harris, Russ. *The Happiness Trap: How to Stop Struggling and Start Living*. Boston, MA: Trumpeter, 2008.

Harris, Russ. *The Illustrated Happiness Trap: How to Stop Struggling and Start Living*. Boston, MA: Trumpeter, 2013. https://thehappinesstrap.com/upimages/Complete_Worksheets_2014.pdf, 2014.

Harvard Medical School. "Treating Social Anxiety Disorder." Harvard Health Publishing, March 2010. https://www.health.harvard.edu/.

Harvard Mental Health Letter. "In Praise of Gratitude." Harvard Health Publications. November 1, 2011. https://www.health.harvard.edu/mind-and-mood/in-praise-of-gratitude.

Hayes, Steven C., and Smith, Spencer. *Get Out of Your Mind and Into Your Life: The New Acceptance and Commitment Therapy.* Oakland, CA: New Harbinger, 2005.

Hayes Steven C., Strosahl K. D, and Wilson K. G. *Acceptance and Commitment Therapy: An Experiential Approach to Behavior Change*. New York, NY: Guilford Press, 1999.

Kabat-Zinn, Jon. *Full Catastrophe Living: Using the Wisdom of Your Body and Mind to Face Stress, Pain, and Illness*. New York, NY: Dell Publishing, 2013.

Kok, Bethany E., et al. "How Positive Emotions Build Physical Health: Perceived Positive Social Connections Account for the Upward Spiral Between Positive Emotions and Vagal Tone." *Psychological Science*, 24(7): 1123–1132, July 1, 2013.

Lieberman, Matthew D. *Social: Why Our Brain Are Wired to Connect*. New York, NY: Crown Publishers, 2013.

Lock, Anna. "Overcoming Social Anxiety Through Assertive Communication." National Social Anxiety Center. https://nationalsocialanxietycenter.com, November 18, 2016.

Martin, Elizabeth I., et al. "The Neurobiology of Anxiety Disorders: Brain Imaging, Genetics, and Psychoneuroendocrinology." *The Psychiatric Clinics of North America*, 32(3): 549–75, 2009.

Mayo Clinic. "Social Anxiety Disorder." Accessed January 3, 2020. www.mayoclinic.org.

Morgan, Nick. "5 Quick Ways to Organize a Presentation." *Forbes*, September 6, 2011.

National Social Anxiety Center. *"CBT Strategies to Overcome Social Anxiety."* Accessed January 20, 2020. https://nationalsocialanxietycenter.com/cognitive-behavioral-therapy/social-anxiety-strategies/.

Neff, Kristin, and Germer, Christopher. *The Mindful Self-Compassion Workbook: A Proven Way to Accept Yourself, Build Inner Strength, and Thrive.* New York, NY: Guilford Press, 2018.

Neaman, Lauren. "Acceptance and Commitment Therapy for Social Anxiety." National Society Anxiety Center, July 16, 2017.

Rosenberg, Marshall. *Nonviolent Communication: A Language of Life.* Encinitas, CA: PuddleDancer Press, 2003.

Safikhangholizadeh, Sima, et al. "The Survey of Effective Communication Skills Training on Anxiety and Social Phobia among Students." Department of Psychology, Ahvaz Branch of Islamic Azad University, Ahvaz, Iran: November 10, 2017.

Shapira, Allison. "Breathing Is the Key to Persuasive Public Speaking." *Harvard Business Review*, June 30, 2015.

Stoddard, Jill A., and Afari, Niloofar. *The Big Book of ACT Metaphors: A Practitioner's Guide to Experiential Exercises and Metaphors in Acceptance and Commitment Therapy.* Oakland, CA: New Harbinger, 2014.

Zetlin, Minda. "11 Graceful Ways to End a Conversation That Work 100 Percent of the Time." *Inc.* Accessed March 20, 2020.

Index

Acknowledgments

I want to deeply thank and wholeheartedly appreciate the following people for their contribution:

My husband, Neil. This book would not have been possible without his innumerable acts of support, love, and encouragement.

My children, Luca and Solenne, who pull for me to be present.

My parents, Kaye and Alex, and siblings, Andrew, Nina, and Leslie, for their love and support.

My dear friends for their encouragement: Rebecca Hamlin, Serena Naramore, Faith Blakeney, Melissa Schilling, Heather Grossmann, Kerri Berkowitz, Brooke Buchanan, Rachel Murray, Brad Bowen, Jess Wilton, Melissa Ciappeta, Suzannah Neufeld, Lindsay West, Kari Sundstrom, Rosa Terrazas, and Daria Mace.

My amazing editors, Lori Tenny and Rick Chillot, who worked with me closely with such patience, support, and helpful guidance. Thank you to everyone at Rockridge Press.

Aaron Beck, Joseph Wolpe, Steven Hayes, Russ Harris, Marshall Rosenberg, Kristin Neff, Albert Ellis, and Jon Kabat-Zinn for their pioneering work and development of the tools that are used throughout the book.

David Burns, for his development and training of TEAM CBT. His trainings, workshops, books, tools, and podcasts have been invaluable to my growth and capacity to help others.

My mentor, Mike Christensen, whose guidance with TEAM CBT over the years has advanced my skills and approach to helping people with breakthroughs.

My instructors at the Feeling Good Institute whose teaching has been transformational, including Jill Levitt, Heather Clague, Rhonda Barovsky, Maor Katz, and Angela Krumm.

Last, a special thank-you goes out to my clients. Their vulnerability, willingness to learn and grow, and courage to face their fears are the inspiration for this book. It's an honor to be on the journey with you, and your breakthroughs are the joy of my work.

About the Author

Alison McKleroy, MA, LMFT, is a psychotherapist, cognitive behavioral therapist, and communication expert with more than 15 years of clinical experience empowering people to be courageous, confident, and in action, creating a life that sparks joy. Her private practice specializes in using effective tools for treating anxiety, social anxiety, and self-esteem.

Alison is a speaker and trainer, leading groups on topics related to the neuroscience of creativity, happiness, and well-being. She is also a coach and founder of Center for Spark, which offers education and training that ignite lasting practices for joy, ease, vitality, and connection.

An avid traveler with an incurable case of wanderlust, Alison has traveled to over 25 countries, has lived and worked on five continents, and speaks four languages. She lives in Northern California with her husband and two children.

Learn more by visiting www.centerforspark.com (courses and retreats) and www.therapyinrockridge.com (clinical practice).